T0388227

Narratives of Faith from the Haiti Earthquake

This book presents an in-depth ethnographic case study carried out in the years following the 2010 Haiti earthquake to present the role of faith beliefs in disaster response. The earthquake is one of the most destructive on record, and the aftermath, including a cholera epidemic and ongoing humanitarian aid, has continued for years following the catastrophe.

Based on dozens of interviews, this book gives primacy to survivors' narratives. It begins by laying out the Haitian context, before presenting an account of the earthquake from survivors' perspectives. It then explores in detail how the earthquake affected the religious, mainly Christian, faith of survivors and how religious faith influenced how they responded to, and are recovering from, the experience. The account is also informed by geoscience and the accompanying "complicating factors." Finally, the Haitian experience highlights the significant role that religious faith can play alongside other learned coping strategies in disaster response and recovery globally.

This book contributes an important case study to an emerging literature in which the influence of both religion and narrative is being recognised. It will be of interest to scholars of any discipline concerned with disaster response, including practical theology, anthropology, psychology, geography, Caribbean studies and earth science. It will also provide a resource for non-governmental organisations.

Roger P. Abbott is Senior Research Associate at The Faraday Institute for Science and Religion, U.K.

Robert S. White, FRS is Professor of Geophysics in the Department of Earth Sciences at the University of Cambridge and Director of The Faraday Institute for Science and Religion, U.K.

Routledge Focus on Religion

Amoris Laetitia and the Spirit of Vatican II
The Source of Controversy
Mariusz Biliniewicz

Muslim and Jew
Origins, Growth, Resentment
Aaron W. Hughes

The Bible and Digital Millennials
David G. Ford, Joshua L. Mann and Peter M. Phillips

The Fourth Secularisation
Autonomy of Individual Lifestyles
Luigi Berzano

Narratives of Faith from the Haiti Earthquake
Religion, Natural Hazards and Disaster Response
Roger P. Abbott and Robert S. White

For more information about this series, please visit: www.routledge.com/Routledge-Focus-on-Religion/book-series/RFR

This scene of nobility in the face of destruction strikes us as iconic of Haitian resolve against defeat.

Source: Photo: Jan Grarup

Narratives of Faith from the Haiti Earthquake

Religion, Natural Hazards and Disaster Response

Roger P. Abbott and Robert S. White

LONDON AND NEW YORK

First published 2019
by Routledge
2 Park Square, Milton Park, Abingdon, Oxon OX14 4RN

and by Routledge
52 Vanderbilt Avenue, New York, NY 10017

Routledge is an imprint of the Taylor & Francis Group, an informa business

British Library Cataloguing-in-Publication Data
A catalogue record for this book is available from the British Library

Library of Congress Cataloging-in-Publication Data
A catalog record for this book has been requested

ISBN: 978-0-367-13406-8 (hbk)
ISBN: 978-0-429-02628-7 (ebk)

Typeset in Times New Roman
by Apex CoVantage, LLC

We dedicate this work to those survivors of the Haiti earthquake who have given their time and energy to sit so patiently and to speak so movingly as they shared with us some of the most awful moments of their lives. We dedicate it in memory of those who died from a disaster that was so unnatural, in the prayerful wish that these narratives will make a contribution to helping people live safely with hazards that are essential aspects of the beautiful world our Creator has made and sustains for us to live in. We dedicate this work to making a contribution to the new narrative Haiti deserves and to declaring Haiti to be a beautiful place.

Contents

Acknowledgements

Carrying out the research for this book has been an enormous privilege and challenge, and we are grateful to The Faraday Institute for Science and Religion, in Cambridge, for the encouragement to engage in the work necessary for such a book as this. We could not have fulfilled the work without generous funding from the Templeton World Charity Foundation (TWCF). Their interest in promoting the science-religion debate has enabled us to carry that debate into the practical sphere of catastrophic disasters, though the views presented here are those of the authors and not necessarily those of TWCF. We also want to acknowledge the assistance provided by our Haitian research assistants who endured with incredible patience the frustrations of working in their own country with foreign (*blan*) researchers. We thank Messrs. Odnel Petit Las and Cedric Bijou for their dedication to the research project; for being such incredible gatekeepers who unlocked so many "doors" for us through their skills in navigation, cultural education, courteous introductions and painstaking translation. We also thank our loyal drivers Jean Louis Edvens and Jean-Eric for enabling us to reach both urban and mountainous destinations safely; their role was critical to our success. The Rev. Dr. David Bidnell gave us invaluable assistance in cultural education prior to arriving in Haiti and used his fluent Haitian *Kreyol* as a translator during our first fieldwork trip. Samaritan's Purse, the international non-governmental organisation, enthusiastically and generously provided us accommodation, office space and vehicle use from their Titanyen and Jax Beach bases. We thank Jack Boothroyd and Routledge, our publishers, for their interest in and guidance through the publishing process. We thank our wives and families for their support for this project and for allowing us the prolonged periods spent away from home that are required for ethnographic fieldwork. Finally, we thank God for the guidance and provision he has made, and for the privilege of working with and meeting so many Haitian survivors of one of the most catastrophic disasters of modern times, for the many lessons they have taught us about faith and disaster.

Introduction

Haiti needs a new narrative

Narratives are hugely powerful. The stories we believe about ourselves and the stories others write about us often define us. In today's media-empowered world, it is those narratives written by others that tend to stick most and, alas, tend to hurt most. This point was not lost on Haitian novelist Roxane Gay in her collection of stories entitled *Ayiti*, stories of people who are "tired of the assumptions, tired of everyone getting their stories wrong." In her review Erin Bartnett comments, "The stories create conversations that travel back and forth between Haiti and America to challenge the over-powering news narratives that undermine the beauty, desire and resilience of the Haitian diaspora."[1]

As the reader will gauge from what they read hereafter, it is our view that Haiti is a Low-Income Country (LIC), among the lowest of the low in fact. This is not from her own choice, but because of the decisions of High-Income Countries (HIC) who prefer to keep her in that position by way of the stories they write about her, their narratives about Haiti. Until recently those narratives could be summed up in one single strapline, "The poorest nation in the Western hemisphere," a mantra you will find in almost anything you read written by HIC Haiti commentators.

The day before the eighth anniversary of the magnitude 7.0 earthquake that catastrophically shook the southern half of the country, the news media reported that the President of the United States of America had an Oval Office conversation with both Democrat and Republican lawmakers over immigration legislation. President Donald Trump was heard to speak of certain Low-Income Countries, including Haiti, as "shithole places."[2] In an age of fake news it is hard to discover whether such reportage is true. Nevertheless such an impression of Haiti would not be without precedent. Assuming the truth of the reports, CNN's Anderson Cooper expressed his immediate condemnation, along with his tears, as he recollected his own experiences of reporting on the earthquake in Haiti and of his love and respect for the Haitian people. Many applauded Cooper's riposte; others saw it as celebrity journalism.[3]

That there are aspects of Haiti that are dirty, smelly and subject to human as well as animal defecation cannot be denied. But if Haiti has more of these problems than many other countries, then that is not because Haitians do not care or do not know any better. An unsanitary sewage system does not equate with being an unsanitary person, just a humiliated one.

That Haiti should be labelled as a "shithole country," as if that were the dominant label, is a disgraceful lie, but a lie that describes what has become the dominant narrative for this and other LICs for too long now. If there is anything – sadly – that should expose and disabuse us of this lie, then it is the response of the Haitian people themselves to the earthquake that shook the country and the people so devastatingly at 4:53 p.m. on January 12th, 2010. In the aftermath of that natural hazard, which exposed so many of the human factors that turned a hazard into a disaster, there was much anguish, but also much beauty. So much beauty in fact that we believe the dominant narrative for Haiti should be along the lines that Haiti is a *beautiful place*, because it is. Our prayer is that after reading this book, every reader will agree with us on this much at least. Certainly, the average Haitian citizen, *Fred Voodoo*, as journalist Amy Wilentz calls them, does not deserve to be defined by an expletive. A new, and counter, narrative is warranted, one that can connect the reality of anguish and beauty.

Figure 0.1 An iconic depiction of Haiti's beauty.

Source: Photo – Roger P. Abbott

Can anguish be really connected with beauty in "a narrative that trains the self to be sufficient to negotiate existence," as Stanley Hauerwas insists we should, "without illusion or deception"?[4] Are anguish and beauty not incompatible except in an illusory world? Hence many novels, plays and blockbuster movies have combined these two features: anguish and beauty *have* to go together, or life would be utterly unbearable to act or to watch, let alone to live. How many books have we read that just exhibited anguish? Only medical textbooks, and those works of the classic nihilists maybe. In the worlds of dreams and fiction we insist that anguish must be accompanied by beauty somewhere along the line and certainly come the end. However, earthquakes are not the products of dreams or fiction: they are for real.

This is a crucial point. Therefore, the beauty we need so desperately in times of anguish must be a true beauty. It cannot afford to be some cheap, superficial stuff that simply deceives us, and others, into thinking we are beautiful, when really we are no such thing. It must not deceive; it must not be an illusion of beauty. It must be the real deal! Anything less and it will shatter those who look to it most – the anguished. The beauty we speak of can also be defined, in theological terms, as hope.

For many in today's increasingly secularised world, hope, as understood in the classical Christian tradition, is often represented as belonging to the superficial stuff. The common strapline, first coined, satirically, by the Swedish born American song-writer Joe Hill, back in 1911, is: "Pie in the sky when you die." At that time Hill was cynically reflecting on the emphasis of the Salvation Army's quaint song lyrics, "*You will eat, bye and bye, In that glorious land above the sky; Work and pray, live on hay, You'll get pie in the sky when you die.*" Hill believed the Sally Army in his day was neglecting a vital social responsibility toward feeding the poor. Whereas there may well be strands of Catholic and Protestant theology and practice in Haiti that are short on addressing seriously the structural poverty of their nation, there is much that combines social and theological integrity, and which forms the mainspring for beautiful hope even in the midst of anguish. As the Haitians say, *Lespwa fe vie* (Hope lets you live).

Though the Haitian people that we spent hours interviewing were also well aware of their needs in this life, and their hopes for this life, such was the magnitude of their anguish that the "oxygen" they survived on was the belief they held passionately, and resolutely, in the eschatological hope promised by God in the gospel of Jesus Christ. Throughout our experience in disaster response work and research, both in the United Kingdom and globally, we have met very few survivors who regarded such a concept of hope as something they felt able to mock. It tends to be the armchair spectators to disasters who do the mocking.[5] The Peruvian liberation theologian Gustavo Gutiérrez reflected, "Working in this world [of the poor] and becoming familiar with it, I came to realise, together with others, the first thing to do is to listen."[6] This is why listening to survivors of the catastrophic earthquake that struck Haiti in 2010 can be significantly different from listening to the spectators of it.

Gracing anguish with hope is not something that happens automatically, not on a societal or on an individual level. It is learned, discipled; it is a process in which people make varying degrees of progress, encounter many falls, surges and regressions. The narratives you will read here exhibit all of these features. However, the important thing, in the end, is that hope graces anguish.

Haiti

So let us introduce you to Haiti, to how we came to be there and to the ethnographic work we conducted there at various times spanning the years 2010–2014.

Sadly, some spectators we spoke to about our research work in Haiti did not even know where Haiti was, or did not even know there was a Haiti, responding to us with words like, "What was it like in *Tahiti* then?" To be honest, we cannot be too hard on people not being excited about Haiti, or for confusing the name with Tahiti even, given that most articles and books about Haiti usually start off by saying that "Haiti is the poorest nation in the Western hemisphere," and then, "It experienced a devastating earthquake in 2010 from which it is still struggling to recover." These stereotypical straplines do not make for great TripAdvisor ratings.[7] If, however, they portray a world away from the reality of Haiti, of deception and illusion, then Haitian anthropologist Gina Athena Ulysse is right when she insists that Haiti needs new narratives.[8] We share that conviction, these ten years after the event.

The only really true, and distinguishing, fact about Haiti amid the stereotypes is the fact that it did experience a devastating earthquake in January 2010. In fact, it was one of *the* most devastating earthquakes when measured by human cost. But then, Haiti also experienced a cruel cholera outbreak of epidemic proportions in the ensuing months of that same year. Then, just what you do *not* need when you are trying to recover from a devastating earthquake and have a cholera epidemic at its peak, tropical storm Thomas dumped huge amounts of rain water onto deforested mountains and valleys in early November 2010.

A baptism of fire!

When Roger first heard of the earthquake that struck the country of Haiti, there was a certain foreboding on his part. Probably like many others involved in disaster response, he imagined not just a large death toll but also a much larger psychological toll of helpless, hopeless and traumatised survivors. To understand the reason for such foreboding you have to try to comprehend that this Caribbean country, covering some 27,750 square kilometres, is also home to 10.3 million people.[9] This means that in 2010 there were roughly 359 people for each square kilometre of the country. However, in the capital, Port-au-Prince, where around 2.3 million people are housed, that density rises to more than 6,000 people per square kilometre, catered for by a seriously dysfunctional infrastructure for sanitation, potable water and power. If that were not risky enough, Port-au-Prince also sits on the main Enriquillo-Plantain Garden geological fault, and though there are some finely constructed houses and commercial structures in this city, these are far outnumbered by poorly constructed business and domestic dwellings, packed densely together, many perched on steep mountainsides or on reclaimed land beside the sea. These constructions, up mountainsides south of the immediate foreshore or in slums actually on the foreshore, made a human disaster time bomb that exploded in conjunction with the earthquake on Tuesday, January 12th, 2010.

There can be no doubting the history of anguish in Haiti, earthquake apart. After all, as Robin Kelley cryptically asked: "Crush a nation's economy, hold it in solitary confinement, and fuel internecine violence, and what do we get?"[10] When people have struggled for most of their lives to get a home of some kind, to find a job that provides at least some basic quality of life to raise some (often, many) children and to enjoy the community found in the Haitian *lakou*, it inspires learning in our view.[11] When that "life" becomes a perpetual struggle against an elitist, State machine that seems intent on crushing people it is supposed to care for; when that "life" is routinely subjected to being squashed by the foreign policies of greedy, rich and largely white nations who have yet to forgive those black slaves back then in 1791–1804 for even daring to attempt gain their independence, let alone succeed in doing so; when you find that amount of "life" crushed even more (sometimes literally) by an earthquake and by the deadly disease of cholera that was introduced so soon afterwards; then it leaves you wondering, how do they cope with this? How do they get up and come back from this? What kind of person can do that (because that kind of person must have a lot to teach us)?[12]

So, our purpose is to attempt to convey both the anguish and the hope from the voices of people we have had the privilege of meeting and listening to as they shared with us details of the most awful moments in their lives, sometimes, ironically, while we met in some of the most beautiful locations. In the words of Amy Wilentz, as she analysed her role as a white Westerner interviewing predominantly poor Haitians: "I'm writing this story for you, reader. I'm bringing their squalid lives into your nice house, your apartment."[13]

Realism and honesty

The Jesuit priest and scholar Jon Sobrino, based in El Salvador, has argued that earthquakes expose aspects of truth that we often prefer to conceal; catastrophes provide an X-ray of the country. In fact, he asserts that "the primary demand of an earthquake, and in the long term the most decisive, is not merely to do acts of mercy but to be honest toward reality."[14] We seek to adhere to that perspective throughout this book; along with the acts of mercy, we shall not spare the reader the reality of what people experienced and why they did so.

Therefore, this book is not intended to be a romantic, or triumphalist, story. It is true; there is something in Haitian life and culture that can be exceedingly romantic. Haitians are romantic people by nature and, frequently, by physique and body language. However, beneath the air of romance there are some dark and depressing realities. What happened after the earthquake revealed these starkly, as every disaster does to its affected people and to others who have eyes to see. When we speak of hope, we do not propose to dilute the anguish one bit. The anguish is historically deep. Nevertheless, so is this beautiful virtue of hope, a beauty that has been taught and honed by anguish and by faith, which resonates with the Old Testament psalms, and with what has made those hymns so endearing and attractive to people who suffer. It is also why the Apostle Paul could write:

> We rejoice in our sufferings, knowing that suffering produces endurance, and endurance produces character, and character produces hope, and hope does not put us to shame, because God's love has been poured into our hearts through the Holy Spirit who has been given to us.
>
> (Rom. 5:3–5)

This is a lesson we in the so-called developed, and often more cosseted, world have largely forgotten. We tend to live by an artificial "beauty," the expensive, manufactured "beauty" of the cosmetic and fashion industry – where what you see on the outside may have little bearing on what can lurk on the inside. Such beauty shuns or, more realistically, masks anguish of any kind – physical, psychological or spiritual. It is not the experience of the Haitian people we met though, young or old. They could neither shun nor mask the earthquake or the history; it was in their face, as we say.

To be frank, Haitians are not immune to superficial brands of hope and beauty, their own social anaesthesia. Haitians love to "party." *Carnival* – that season leading up to Lent, of almost continuous "partying" with street parades, delicious *Kreyòl* food and intoxicating *klerin* (Haitian rum)-fuelled, dance – is an annual date for the State to encourage *Fred Voodoo* to try and forget his or her anguish and to celebrate and enjoy their cultural beauty. *Carnival* is also an occasion when the State often drains the nation's coffers of far more than can actually be afforded, thus depriving areas of Haitian life that desperately need finance due to their critical state.

However, Haitians know very well that after *Carnival* is over the root causes of their anguish will still be there through another year. It is just the brief relief that is sweet and, maybe, even necessary, given the persistent anguish at large in the nation. When Stanley Hauerwas wrote: "The kind of character the Christian seeks to develop is a correlative of a narrative that trains the self to be sufficient to negotiate existence without illusion or deception," he said something profound.[15] That is what our research in Haiti has shown us. But it is also something profoundly disturbing when we think of how so many of our cultural equivalents back home in the West are ill-prepared for what the Haitians experienced in 2010. That is simply because illusion and deception are endemic in our pampered culture. *Carnival* aside, this was rarely so with the Haitians we met, many of whom saw even *Carnival* as a social anaesthetic they would not resort to themselves as Christians; they had other ways to sustain hope and beauty.

Some would say that the beautiful hope of the Haitian spirit can be heard most, and experienced most, in the churches. For it is often in their meetings for prayer and worship and ministry of the Scriptures, that Haitians express themselves and sing their purest and most melodic strains. Again though, we want to be careful not to triumphalise that claim or to romanticise the state of the churches in Haiti or to suggest that melodic strains are confined to the churches. There can be too much gossip, discordance and even corruption in the churches; too much hand-in-glove with political and commercial elites; too much fixation with power and material prosperity in the name of Jesus; too much "empire building" by people happy to use religion for exploitative financial gain. Nevertheless, what has affected us deeply at times is that it is often in church contexts that both the anguish and the hope can be seen and felt most beautifully.

What Roger actually experienced from those initial deployments in Haiti during 2010 and 2011 was quite different from what he had expected to find regarding the Haitian people who had been hit by that "double whammy" of the earthquake and the killer disease of cholera. When he took the evening flight out of Port-au-Prince in 2011, he could not help wondering why there was such a difference between what he had expected and what he had actually found. What he found was a people strong in a hope born of their faith, who kept on praying, praising, smiling and surviving in the wake of tragedy and suffering. He prayed that it might be possible to return one day and research the answer to that question. What follows is the fruit of that prayer being wonderfully answered. This is the route we shall take.

Road map

Chapter 1 gives an account of the earthquake from survivors' perspectives. It presents their at times shocking narrative evidence experienced in the worst earthquake affected areas in Haiti. This chapter narrates what survivors heard, saw and smelled. Though harrowing in places, we think it is important to introduce readers to the realities of what it was like to have lived through this terrible earthquake. The chapter attempts to give a physical background to the extreme circumstances with which survivors' faith had to cope.

Chapter 2 sets out the psychological and emotional impacts arising out of the survivors' experiences, and the influences of their culture and faith. In particular, it focuses on how survivors witnessed deaths and injuries as they attempted to navigate the ensuing chaos and destruction in the immediate aftermath of the earthquake. They were exposed to the death of children; to having no chance to say "Goodbye" to loved ones; to seeing and hearing the dying; to the smell of death; to worshipping amid the dying; and to the system of body disposal. All of these factors contributed to the stress with which survivors had to cope.

Chapter 3 explores how the earthquake affected the religious faith of survivors and how religious faith (mainly Christian) influenced how they responded to, and are recovering from, their physical and psychological experience of the earthquake. The influence of themes such as the Bible, creation, providence and eschatological hope is described in terms of the positive and negative contributions these made towards their disaster recovery and to the future mitigation of disasters. A response is also proffered that counters the common Western assumption of fatalism being a typical Haitian religious outlook that hinders disaster recovery and mitigation. We also comment on the evidence of the negative impact of Western religious influences throughout Haitian history and suggest how Haitians resisted and coped with such influences, thereby developing, by now, culturally embedded coping strategies.

Chapter 4 reflects on the science that was available (or not) to the Haitian people and how it could have been used to reduce disaster risk significantly in the earthquake-prone areas of the country. The level of earthquake awareness across the demographic spectrum reflected by survivors was, relative to the science available to their Government institutions, appallingly poor and played a significant role in the resulting scale of human and material damage. On the other hand, what level of education there had been played a significant role in providing a degree of pastoral care and assurance to survivors, contributing to their limited resilience. In particular, such factors as a "criminal" lack of education regarding seismology and disaster preparedness are highlighted to show how a natural hazard could have been reduced in its disastrous effects through access to good education, an access denied to many of the poor in Low-Income Countries like Haiti.

Chapter 5 sets out how additional, more historically embedded, "complicating factors" within Haiti have created a culture of constant struggle for the majority of Haitian people, which, ironically, has also helped to create a culture of survivability among Haitians. That exposure to stressors on a regular basis has contributed, so we argue, to the survivors' capacity for coping with and recovering from the trauma of a catastrophic event, and that capacity can be enhanced by the contribution of faith.

Chapter 6 draws together lessons from our research of the Haitian experience for more global contexts of disaster response and recovery. We highlight the significant role that religious faith can play in exhibiting and disseminating these lessons through networks of churches and faith-based organisations globally, in a bid to ensure that the human factors of "natural" disasters are addressed with social integrity and equity. Typical perspectives that frequently frustrate effectiveness in this role are challenged.

Notes

1 Erin Bartnett, "Roxane Gay on the Trauma and Triumph of the Haitian Diaspora," *Electricliterature.com*. Online: https://electricliterature.com/roxane-gay-on-the-trauma-and-triumph-of-the-haitian-diaspora-12491de1b3e0. Accessed: 28/08/2018.
2 Josh Dawsey, "Trump Derides Protections for Immigrants from 'Shithole' Countries'," *The Washington Post*, January 12, 2018.
3 Amy Wilentz, *Farewell Fred Voodoo: A Letter from Haiti* (New York: Simon & Schuster, 2013), 59.
4 Stanley Hauerwas, *A Community of Character: Towards a Constructive Social Ethic* (Notre Dame: University of Notre Dame Press, 1981), 132.
5 We find it significant that in moments of individual, or even mass, tragedy people will immediately construct local shrines where they lay flowers, toys and messages that often express hopes concerning the casualties' lives after death. See Anne Eyre, "In Remembrance: Post Disaster Rituals and Symbols," *Australian Journal of Emergency Management* 14 (Spring 1999): 23–9.
6 Gustavo Guttiérrez, *The Density of the Present: Selected Writings* (Maryknoll, NY: Orbis, 1998), 171; Michael Griffin and Jennie Weiss Block, eds., *In the Company of the Poor: Conversations with Dr. Paul Farmer and Fr. Gustavo Guttiérrez* (Maryknoll, NY: Orbis, 2013), 23.
7 In fact, TripAdvisor regularly recommends the cruise ship line of Royal Caribbean with its resort of Port Labadee on the north coast of Haiti. Its economic benefit to Haiti from tourism is minimal. However, as one commentator has said: "To be completely honest, the distance between Labadee and Port au Prince is . . . 130 miles. And, as far as the average Haitian is concerned, it might as well be on the moon. Labadee's little piece of paradise is completely isolated

from the rest of the island, and its 230 workers are carefully culled from the millions of Haiti's citizens." (Bruce Watson, "Cruise Line Visit to Haiti Highlights Ugly Side of Paradise," *Daily Finance*, January 22, 2010).

8 Gina Athena Ulysse, *Why Haiti Needs New Narratives: A Post-Quake Chronicle* (Middletown, CN: Wesleyan University Press, 2015).

9 Haiti's land area is slightly less than that of Maryland, U.S., but with twice the population. Haiti is less than half the area of the Republic of Ireland, but has over double the population. For population figures see *Knoema: World Data Atlas*. Online: https://knoema.com/atlas/Haiti/Population-density. Accessed: 11/11/2018.

10 Robin D.G. Kelley, in the Foreword to Ulysse, *Why Haiti Needs New Narratives*, xiv.

11 The *lakou* (the *Kreyòl* means space, or a large courtyard) refers to the open space around which Haitians traditionally build their homes. The families are often inter-related, but even in instances of this not being the case, the residents look out for each other.

12 See Westenley Alcenat, "Reason in Revolt: The Case for Haitian Reparations," *Jacobin*, April 1, 2017. Online: www.jacobinmag.com/2017/01/haiti-reparations-france-slavery-colonialism-debt/. Accessed: 14/01/2017.

13 Wilentz, *Farewell Fred Voodoo*, 204.

14 Jon Sobrino, *Where Is God? Earthquake, Terrorism, Barbarity*, trans. Margaret Wilde (New York: Orbis, 2006), 12–13, 16, 30.

15 Hauerwas, *A Community of Character*, 132.

1 When our earth shook

Jou trambleman tè a.

(Kreyòl refrain by interviewees: The day of the earthquake)

We have said that to understand the reality of an earthquake, and to react appropriately to it, we must "let ourselves be affected" by the tragedy.

(Fr. Jon Sobrino, Jesuit priest, El Salvador)[1]

Introduction

It is hard for those who hear about a tragedy second-hand to be affected by it, however moving the first images and reports may be. Jon Sobrino suggests that this failure to be affected is also why tragedies are so quickly forgotten.[2] As *blan*[3] researchers, definitely not survivors, of the Haiti earthquake, we feel the weight of Sobrino's words. Nevertheless, as we have been privileged to sit with survivors as they have shared with us the worst moments in their lives, in many cases begging us to tell their experience to the world outside Haiti, we feel obligated to share with readers what they have told us so that the Haiti earthquake tragedy is not quickly forgotten.[4]

At 4:53 p.m. on Tuesday, January 12th, 2010, the magnitude 7.0 seismic shock struck. It lasted 35 seconds, continued rumbling through dozens of aftershocks over the following two weeks and was felt most strongly in the West and South-West departments (regions) of the country.

It was in those departments where the majority of structural and infrastructural damage occurred, and where most lives were lost and life-changing injuries inflicted. We selected participants from different demographic areas: from the Cité Soleil slum area, on the reclaimed land of the foreshore of the capital, Port-au-Prince; from the central area of the capital; from the outlying cities of Léogâne and Petit Goâve, around 30 and 50 miles respectively, west of the capital; and from the rural mountain regions of Durisy,

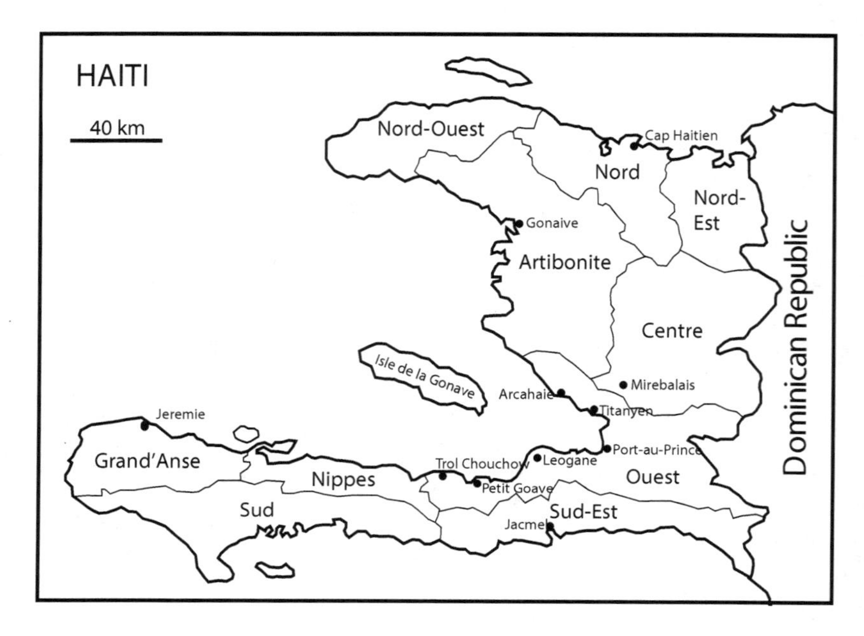

Figure 1.1 Haiti's administrative departments, showing the main centres mentioned in the book.

which overlooks Petit Goâve, and Trol Chouchow, mountains that rise up along the coast to the west of Petit Goâve.

Each of these areas, together with the southern city of Jacmel, experienced heavy tremors and extensive damage from the earthquake.[5] They represent the worst earthquake affected areas in Haiti. Fewer people in the rural areas died, or were injured, although some rural residents, and especially children who were attending school in the cities when the earthquake struck, died under the rubble of city buildings or were injured by debris.

Where people happened to be when the earthquake struck was random, not in any statistical sense but in the everyday sense of the word: happenstance if you like. It could appear that who died, who was injured and who survived unscathed was all a matter of being in the wrong/right place at the wrong/right time, no more than that. That is also what journalist and writer Amy Wilentz seems to conclude in her powerful book *Farewell Fred Voodoo*, written after the earthquake.[6] It was just chance, unless you were Haitian that is. For Haitians also find "accidents" and "randomness" hard to accept. For the Haitian there must be an explanation, a reason, which is where religion – Vodou, Catholic or Protestant is often a valuable hermeneutical tool for them.

Wherever our participants were, either they nearly died or they witnessed others dying. The common *Kreyòl* refrain, "*Jou trambleman tè a*" (The day of the earthquake), which participants used in our interviews, revealed that this dreadful day was now embedded in their minds as we asked them to describe for us what it was like – what they heard, saw, smelt and felt – in as much detail as they felt able.

Omens?

An essential scientific fact to understanding an earthquake is that earthquakes do not generally give warnings, unlike the hurricanes that assault Haiti annually and which can be tracked for days before landfall. Even sporadic earth tremors do not necessarily augur a major earthquake. Many tremors have been felt along the Enriquillo-Plantain Garden fault since the last earthquake in 1770 without anything more serious occurring until the 2010 event.[7] By all accounts, *Jou trambleman tè a* started out as so many other days in Haiti at that time of year: sunny, hot and bright, with no hints of the natural hazard that was to strike so briefly, yet with such devastating consequences, later in the afternoon.[8] People got up to another beautiful sunny, January morning, about the coolest, or least hot and wet, part of the day, and of the year, for Haitians. People went off to their work, if they had any, or to school, college or university, if they enjoyed that much privilege; or they remained at home and mooched around.[9] None of our participants was worrying about an earthquake happening during the course of that day. In Haiti, there are plenty of other things to worry about on a daily basis, like getting a single meal for the family, getting work, accessing education for the children, purchasing some paracetamol to help reduce a fever and a myriad other things we may take for granted. In our research journal, in 2013, we summed up the impression of that day as related to us by a group of church leaders we met in the provincial city of Petit Goâve. We noted, "The day had been very normal: sunny and warm. Yet they spoke of it as a day like never before!" In Grand Goâve, a smaller, nearby town, Marcos told us that what struck him about that day was that once the dust had settled, following the initial shock, the sun was still shining and the birds started chirping as they had been right up until the earthquake happened. In fact, the only remote hints of unusualness on that day were those related to us by Micheline, Henri and Pierre.

At her home in Grand Goâve, as she prepared to go to work that morning, Micheline thought that the atmosphere, for no reason, seemed "sad, as if something was going to happen." As Henri sat in class in his school in

Delmas, Port-au-Prince, at around 10 a.m. that morning he and the other students heard the glass in the windows rattling, this kept on recurring throughout the morning. Pierre was also living in Port-au-Prince at the time. He had been thinking of looking for work, and so he was in and out of home at various points throughout the day. He was also very nervous because of a dream he had the previous day, warning him of some terrible impending event soon to happen. Micheline never so much as thought the atmosphere's sad feeling signalled a devastating earthquake. Henri had no reason to think the rattling jalousies meant anything other than more heavy goods vehicles or tractors than usual were passing. In the end, Pierre went to his neighbour to have some food and drink and to relax. This is when the ground started to shake violently. Pierre and his neighbour went out into the street and got bounced around with the vibrations of the earth. The ground ripped apart not far in front of them and buildings began to collapse. Pierre's children were taking a shower in the outside bathroom, and they just managed to escape before the cement block cubicle collapsed. In other words, any possible relationship either of these experiences bore as possible omens of the coming earthquake were just speculative and recalled with the perspective of hindsight.

Sounds, tremors and waves

It is worth noting that some phenomena were common throughout the areas. These related to certain sounds, to the tremors and wave motions participants heard and felt.

For those of us who have never lived through an earthquake it can seem difficult to imagine what the physical experience must be like.[10] Indeed, most Haitians, few of whom had ever experienced an earthquake before, struggled to understand what was happening when the earth shook, cracked or heaved beneath them.[11] One aspect that may assist our imaginations in understanding can be the sound the earthquake made: that alone can instil terror. People said they would never forget those sounds. In fact, Haitians made up an onomatopoeic term to describe them, *goudoo-goudoo*, repeated rapidly. The term for some was a therapeutic tool, which substituted for them having to say the actual words for earthquake (*trambleman tè a*), lest just saying the word might presage another earthquake.[12]

People spoke of a loud noise coming towards them travelling from east to west.[13] Some described the noise being like a terrifying, deep, growling sound; one survivor described it being like that of a dragon! For some it reminded them of a helicopter or aeroplane taking off or crashing. Others likened it to loud thunder or to a loud wind blowing. A few likened it to an electricity surge, a common feature for Haitians. One Port-au-Prince participant said it sounded like a loud ship's horn. Most said it sounded just like the noise of heavy goods vehicles or the caterpillar-tracked vehicles passing by that they were more routinely familiar with, only this time it was much louder and more persistent. However, whatever the noise sounded like, the surprise or terror it inspired was nothing compared to that brought on by the waves and tremors felt through the ground and the structures that lurched sickeningly, then, too often, collapsed.

The tremors seemed to vary in intensity, according to the individual's experience. There were no light tremors; all who felt the shaking felt it to be violent. Often it would immobilise people: if they sought to run, they fell over or were spread-eagled on the floor. For some, there were also waves of motion, as if the land moved in sickening swells, like a rolling sea. In the mountain regions especially, we found people complaining of shaking in their limbs and muscles, even three years after the event, perhaps indicative of the violence of the quake, or at least, of the neurological impact.

Location

Traumatic events seem to leave a cognitive imprint in the memory regarding location.[14] Participants' recall of where they were was generally vivid even three years after the earthquake. They were in a variety of places, a result of routine living and indicative of their unawareness of exposure to risk. This was the time of day for transition. As the academic day was divided up into two sessions – a morning and an afternoon, with some institutions running evening classes as well – many students and staff were in or around school premises at the time. Others had already left and were én route for home, on foot, in *motos* and *tap-taps*. Workers who had finished their day shifts were returning home; those on early shifts and the many who make up the vast unemployed sector were already home or *hangin' out* on the streets or, in the rural regions, sitting in the shade of a tree, playing dominoes and drinking *klerin*. Neighbours, especially young people, were gathered in homes where there was a TV set, ready to catch the next episode of a popular soap opera showing around that time on a Tuesday. Some of our participants were in church. Churches often had prayer-meetings arranged for around 5 p.m., so congregants could drop in on their way home from work. Some people were travelling about in their own vehicles.

It is impossible to describe every participant's experience, so we have selected individual cases that, along with others, we follow through the course of this book. They are not statistically representative, but, from our observational studies, they are not unique.

Port-au-Prince[15]

The capital city of Haiti, Port-au-Prince, is a massive urban sprawl, built on a low-lying coastal foreshore before the plain gives way to an increasing ascent as one travels inland, up the steep mountainsides. The city is a mix of slums, shanty-towns, Government buildings, hotels and gated residential properties. It is home for some 3 million inhabitants, but a population that swells massively when the additional *machanns*[16] descend upon the city from the surrounding countryside for commerce and livelihoods. The city has reached the size it has by way of the centralising urbanisation that successive governments (including foreign governments and foreign commercial interests) have encouraged, luring the majority rural population to give up their traditional preference for agriculture for the illusory guarantee of a stable job in the *blans'* factories and outsourced businesses in the cities. The Haitian Government and, more significantly, the tiny yet very powerful elite class in Haiti, have manipulated this policy in their own interests. The result has been the creation of a number of slum and shanty-town areas. Most, but not all, are at the edge of the sea on reclaimed land, barely a metre above sea-level, over which the storm-water that frequently deluges the mountains surges back from mountains to sea. Behind these slums lies the city centre, and rising up behind the centre are areas of residential housing, hotels, embassies and administrative buildings reaching to the more exclusive neighbourhoods of Petionville and beyond.

Some people were at home

In Haiti, apart from anything else earthquakes test intuitive thoughts and assumptions about feeling safe and secure at home. Earthquakes expose how fragile houses are and how fragile people are. In Port-au-Prince the earthquake sent people running in all directions. Some ran out of buildings into the narrow corridors and rabbit-runs between houses only to be crushed by falling masonry; some ran into the streets only to be confronted with electricity wires left hanging or left dragging on the road, electrocuting anyone who happened to inadvertently touch them. Traffic was thrown into chaos, with vehicles smashing into one another or running down pedestrians escaping out of buildings. Some people ran into their houses, even onto the roofs! When you do not know what is happening, blind instinct can take over, even when it might be the wrong thing to do; you may rush into death rather than away from it. Sometimes whatever course of action you adopt seems a lottery as to how it will end up for you.

This was the case for Jonassaint, his wife Dominique and their family of three children. They were living in a third floor apartment in Carrefour at the time of the earthquake. Jonassaint left his work at around 4 p.m. and came home to find his 9-year-old twin boys waiting for him at the gate as usual. Dominique had already arrived back home from her work as a *machann*. Just as the father and sons arrived at the third floor apartment, the whole building started to shake violently. The two boys let go of their father's clothes and, terrified, fled down the internal stairs just as the whole building collapsed, taking Jonassaint, his young daughter and Dominique with it and burying them alive beneath the rubble. Dominique, still conscious, could recall the experience in terrible detail:

> I was in the place with them in the house. Before the earthquake happened, I had just gotten home. I was going to check the kid's uniforms. My habit was to check them every afternoon, wash them and prepare it for the next day so they could go to school the next morning. As I was inside, I heard the chinaware rattling inside. I was asking myself what was going on. My husband was on the balcony with the kids. When I joined them I saw [heard] the noise occurring. It was that same time when my husband and daughter ran for the inside. The twins had already left. I stood still, I couldn't run. Because of the noise that I heard, I saw that the house was going to fall on me, and I cried out "Jesus, are you really going to let me die?" I was in a corridor in the house, with walls to my right and left, and said again "Jesus are you *really* going to let me die?" I saw the house still bending downward toward me. "Oh Jesus! Are you *really* going to let me die?" I then saw something all white that blinded me; next thing I knew I was down on the floor lying down. As I was lying down, I saw the house (apartment) came down, and I was trapped under blocks. I saw my hands weren't injured. Neither did I feel any pain in my ears; after seeing the blood, I noticed that I was cut in the ear. I felt my feet were stuck, but I couldn't get them unstuck. I was lying face down on my breast trying to get free. Afterwards, when our daughter who was in the street, came, and having seen the house destroyed, she thought we were dead. And because she thought we were dead, she fainted in front of the gates. God gave her strength, and made her get up and climb on top of the rubbles. My husband was yelling for people to come and help us . . .

She went on to recall how the ceiling collapsed on them,

> . . . but we don't know how we survived. The ceiling collapsed in a cone shape, and we were under it. And this was our experience with

> God, because we were asking ourselves how is it we survived. When the ceiling collapsed, it collapsed in a cone shape and only our heads were visible to the outside world.

Under the guidance of their other daughter, who was not in the building when it collapsed, rescuers eventually were able to locate and release Jonassaint, Dominique and their daughter. Sadly, the twins were later confirmed as dead, though the family never recovered their bodies.

Some people were still at work

At the Hotel Montana, a classy, up-market complex, built on seven levels, in the Delmas district of Port-au-Prince, a proprietor, Estelle, was outside when she heard a dreadful sound of rumbling in the earth: "It's as if dragons from nowhere had appeared. . . . It is a terrible, terrible sound! It's the innards of the, of the earth, growling!" she described. It was extremely loud, a sound that continued to haunt her still. Then came the shaking, so fierce she could not remain standing "because the soil was jumping up and you couldn't walk." Estelle did not see any buildings actually fall, but when the noise and shaking stopped, there was a huge cloud of dust, which, when it

Figure 1.2 Collapse of Hotel Montana, originally seven storeys.

Source: Photo – Logan Abassi, United Nations Development Programme, CC BY 2.0

cleared, revealed a scene of horrific devastation. The famous hotel, a popular venue for many visiting foreign dignitaries and business people, had almost all collapsed. Estelle's sister, Lucie, meanwhile, was inside the hotel when that horrible noise started. She had felt very uncomfortable and started to run into an office, a designated "safe room," with iron security grills on the window, built to provide security from rioters and intruders. There, as the complex collapsed, she was entombed beneath concrete, rebar, dust and dirt for five days before being rescued. Though seriously injured, she survived by having access to fresh air and by drinking her own urine, as trapped colleagues near to her, one by one, suffocated and died.

Some people were travelling home from work

At 4.53 p.m., the main streets of the capital were gridlocked with traffic taking people home from work. Claire was travelling in a *tap-tap* near the football stadium in central Port-au-Prince when the car started to shake and became very unstable; everyone in the car was calling out "Jesus!" She got out of the car when the tremors had stopped and saw buildings collapsing, and people nursing lost limbs and other terrible injuries. She also witnessed employees at a bank jumping from the second floor in their panic to get out of the building. Her initial thought was to ask herself if this was a civil war, or another *cout d'état*, because she did not understand what was happening. She made her way home on foot to Carrefour, a journey of some 6 miles. At home, she found a scene of terrible devastation. A house at the rear had collapsed onto hers, destroying both houses. Inside the ruin lay her eldest daughter and that daughter's baby, Claire's only grandchild. Claire pleaded with a man to go inside the ruin to look for her daughter and the child, but he refused because the ruin was so unstable. Eventually that night, rescuers were able to get the daughter out alive, but the baby boy, alas, was dead. While walking home, Claire had passed a group of students, bathed in blood, coming from a school. Seeing them made Claire very anxious for Grace, her youngest daughter, who was a student at a nearby school. She desperately wanted to find her.

Some people were at, or returning from, school

Grace said she was just outside her school, in the street, when she heard a noise like a helicopter or plane crashing. In her terror from the noise and shaking, she clung onto her friend and they both walked up the middle of the street, away from the school. During their walk, the two friends had thought of calling into another school on the way but, thankfully, they had decided not to, because, later, they learned that school had collapsed soon after they had moved on. Along the road, as she walked in a trance of shock, Grace witnessed houses that fell down and there was even a foot she saw sticking out of the rubble and dangling in the air. She found her way into the company of some Jehovah's Witness friends who took her under their care and into a safe space where other survivors were gathering, then later that evening she was re-united with her mother.

Emilie, a university student nurse, had returned to her home in the Bourdon district, near the centre of the capital. She and two classmates then returned to the nearby Champs de Mars, a popular gathering place for students in central Port-au-Prince. They went there to revise together for an impending exam. Suddenly, they became aware of many people running, so they also ran, though not realising what was happening, until, that is, someone told them an earthquake was happening and that they should find a place to shelter. At the same time, she noticed that the iconic White House, the Presidential building, erected by the U.S. Marines during their occupation of Haiti (1915–1934), had already collapsed, along with other Government offices. Emilie made her own way back to where her second floor apartment in Bourdon had been, only to discover the whole building had totally collapsed, killing her mother and older sister, whose remains she was not able to recover and hold a symbolic funeral for until the following May.

Léogâne

Léogâne lies some 25 miles west of the capital. It is the ancient cultural, Vodou capital of Haiti. The main Enriquillo-Plantain Garden fault runs through the range of mountains that rise up farther inland from the city, but a fault probably responsible for the 2010 event runs beneath the city with its sprawling population of 134,000. Léogâne also lies on a low-lying coastal delta with a high water table. Some 80–90% of buildings suffered in the earthquake, with a massive human toll as well.

Some people were at home

Virgil, a civil engineer and entrepreneur, had built a hotel complex of some repute in Léogâne city. On the day of the earthquake, he was in his hotel, listening to the news on the radio, when he fell asleep. The sound of the earthquake woke him up. He first resigned himself to the fact that this was to be the moment of his death, but then something in his head told him he had a chance to survive, whereupon he leapt up and jumped out of a window, just before the whole building collapsed.

People at work

Patrick had recently graduated as a young teacher at a college in Léogâne. He had come into school that afternoon expecting to teach a class. When he arrived, for some reason there was no one in the class. On his way home, he met some of his students and tried to encourage them to return to college. However, they showed no interest, so he went to a cyber cafe with a friend. In the same building there were English and Informatics classes going on upstairs. When the walls began to shake he noticed students running down the stairs to exit the building, but he did not understand what was going on. When his friend got up to leave, Patrick decided to go with him just as the boundary wall of the place collapsed. Across from the cafe, there was a Catholic school, and that building with its fence also collapsed. On walking to his home, Patrick felt he was dreaming. A mother was calling out for help to release her son who was trapped beneath her collapsed house, and there was a school he used to go to, with two storeys that had collapsed with many children calling from beneath the rubble.

Some were in prayer meetings

Marion was attending a prayer service in a house near where she lived in Léogâne. At the time, they were praying especially for a young child who kept getting sick. Some folk were sitting inside the house and some, including Marion, were seated on the gallery outside. When the building began to shake terribly, she quickly jumped off the gallery as the house collapsed, killing eight of those inside and injuring others. She managed to get outside the yard and saw buildings falling and a black cloud forming an eclipse effect with the sun. Then she went back into the yard to help lift a concrete beam off a woman's leg. Jean-Pierre, an already disabled man, was also at this prayer service. Because he could not negotiate the stairs, he had a chair to sit on in the gallery outside. As the ground started to shake and people were running he got up to run as well but fell down. He stayed seated on the floor and watched the house "pancake" straight down, just missing him. His face became covered in dust so he could not see until a woman from the church wiped his face and eyes. Then he could see the immense damage that had been caused, and he also heard that his own house had been destroyed.

Out and about/visiting others

At the time of our interview, Bob was Police Commissioner in Léogâne, working out of a series of portakabins that were the temporary police station.

The earthquake had put four out of five police stations in the district out of use. On the day of the earthquake, Bob was walking in downtown Léogâne. When the ground started to shake, he assumed it was a caterpillar-tracked vehicle of some sort passing by. He admitted that being a police officer his next thought was that some kind of bombardment must be happening. He was standing next to a wall, which then began to fall towards him; in running away, he fell down and then saw two houses collapse nearby. He still did not know what was happening. After the shaking stopped, people who knew him, who were pulling casualties out from under the rubble, called out for him to help them. Then he understood it had been an earthquake. Though in a state of shock himself, he began collecting the injured in his pickup truck and ferrying them to hospital, a task made difficult by the roads being blocked. As people died in his pickup while he was driving along, he would replace them with more of the injured before taking them to hospital.

As he proudly showed us photographs of his home, Leroi told us that the finest house in all Léogâne was the one he had lost to the earthquake. We sat on some lawn, just off the ceramic tiled ground floor, literally the floor, that is, all that was left of the once majestic structure that had been home for him and his wife and for their precious twin girls who were products "of years of trying." Leroi was an up-market funeral director. He had owned, and now lost, four houses by the earthquake. He was away from his own home and was in the rear of his father's three-storey house when the earthquake struck. All three storeys came down on top of him until the floor above him came to rest temporarily just 4 or 5 inches above his head. Seeing some daylight, he managed to crawl out before the whole structure fell down flat. As the property slid over, however, it crushed and killed one of his twin daughters.

Petit Goâve

Petit Goâve is a coastal city 46 miles west of Port-au-Prince. The main geological fault-line runs from the headland to the north-east, across the bay in front of the city toward the shore to the west. The city suffers from a huge political disjuncture from the capital and is regarded by the central Government as politically and socially volatile. During our visits to Haiti, we heard of numerous riots in Petit Goâve, mostly due to the routinely inconsistent electricity supply there causing immense frustration to the populace. Up until October 2017, the city hosted a United Nations Stabilisation Mission in Haiti (MINUSTAH) base, which is significant in itself of the civil unrest the Government felt it needed to control in the area. Haitians told me that Petit Goâve was not somewhere the Government officials like to venture as it was a haven of different criminal cartels vying with each other for power.

In the area where most of our interviews were conducted, down by the seashore, the damage was catastrophic, as the following narratives portray.

Some people were at home

Raimond, a community youth leader, was living in what had been originally an Internally Displaced Persons (IDP) camp, at Cupidon, a steep promontory overlooking the Route Nationale 2 on the western edge of Petit Goâve. That day Raimon had just returned from work. His wife needed some more rice for the evening meal, so Raimond gave some money to one of his daughters and sent her and her brother to get the rice from a nearby store across the street. Two minutes or so after the children had left the house, while Raimon and his youngest daughter were indoors, the house began to shake, and the walls of the room began to lean inwards towards them. Raimon and his daughter quickly ran out of the house while his wife ran across the street to fetch the other two children. In fact, she had to stop halfway there because the ground was shaking so violently. When she arrived at the store, she found it partially collapsed with her children buried beneath the rubble. She returned to tell Raimon about the two children trapped beneath the rubble of the store. They both ran back to the site and, with help from others, they dug into the rubble and found the two children alive but trapped by fallen concrete blocks and debris. After persevering with the debris clearance, they were able to release the two children relatively unscathed. Shortly after the children got clear, however, what remained of the building collapsed entirely.

Raphael's experience of the earthquake is extraordinary, though probably by no means unique. It illustrates the apparent randomness of earthquake-related disasters, about who lives and who dies. Raphael's wife had died before the earthquake, leaving him with two children. By the time of the earthquake he was living with a partner with whom he had three more children. That particular day he was heading for the kitchen in his house, to get juice for the three children who were with him. He heard a loud growling noise as the concrete house collapsed around him. "By the grace of God," he said, "I survived." Initially, he assumed that all the children who had been in the house must have surely died. Remarkably, he managed to crawl out from beneath the rubble, unhurt. Altogether, there had been six people in the house when the earthquake struck, and he was the first one to get out. In shock, he could do little to help the others, so neighbours came and dug into the collapsed property. It was they who found Raphael's 19-year-old niece alive and removed her from the debris. They also found the niece's young baby, and the baby's father, both of whom were dead.

Raphael had assumed that at least one of his sons had managed to escape the house in time. He went around the locality looking for him but could not find him anywhere. Still in shock, the following day the desperate father went to the local MINUSTAH base and pleaded with staff to bring their heavy-duty equipment so they could try and see if the son was still under the rubble. When the staff arrived with their equipment and had started to dig, that is when they found Raphael's 16-year-old son, dead. The news broke the father's heart.

Devastated by the loss of loved ones, Raphael left the smashed city of Petit Goâve and went to live with his parents in the nearby province. He knew for sure that the eldest son was dead, and he assumed the youngest, a 6-year-old, had perished under the rubble as well. For some days following the initial search of the collapsed building, having found three dead bodies already, the consensus view was that all who had not been discovered as yet must have died too. Therefore, the search for survivors had stopped. That is, until seven days later, when a rescue team came with their heavy-duty equipment and commenced another search.

Extraordinarily, the remainder of this remarkable story is told from the account of a 6-year-old boy who survived the weeklong ordeal of being buried under the ruins of his home, Raphael's youngest son! The boy had survived the collapse, and from his tomb had been calling out for help. Being immersed in dust, with little water or food, his voice had soon become so weak he was beyond being heard even if anyone had been listening for him.

On one occasion, the boy did spot a rescue worker peering through a hole in the rubble. However, the young boy was so frightened by the stranger that he took a cushion and blocked the hole! Then came the noise of a backhoe being operated to clear the rubble. It got nearer and nearer to where the boy was trapped without knowing the boy was there. All of a sudden, the driver stopped the vehicle, got out and snapped angrily that he was now fed up with doing such miserable work and he was going home. Raphael was there when that happened. He said that he and the rest of the neighbours then went back to their shelters and went to sleep for the night, with no thoughts of any survivors being under the ruins.

On the eighth day after the earthquake, Raphael's brother and sister came to search in the ruins of the house, looking for some products that they had kept in a little store attached to the building. As they were digging, the little boy heard them. He found an aluminium plate and banged on it with a rock. The brother beckoned to his sister and they stopped their digging to listen for the metallic noise. They heard it and began digging again. When they came to the hole the workman had peered through the previous day, the young boy could see his uncle and called out as best he could with his dried up voice, "It's me! I'm here!" Raphael's brother then called out that the young boy was there. Raphael assumed that he meant he had found the boy's dead body, so he was about to go away so he would not see the body when it was recovered. He was assured that the young boy was very much alive, and then everyone available pitched in to remove the rubble with whatever tools they could lay hands on until, eventually, the young lad was released to the overwhelming joy of his father.

Some people were out and about/visiting

Born and raised in Petit Goâve, Martin was 23 years old when the earthquake struck. He had visited a store to buy food. He had then helped his older brother who had been cutting rice that day store away a bag of it in his brother's house. He was passing through a hallway in the house when the ground started to shake. He looked up and saw the floor above coming down on him. He then lost consciousness. He himself cannot recall anything more of what happened. All he knows, from what others have told him, is that some people brought his injured, swollen body out from beneath the rubble of that house and to hospital in Petit Goâve, where he was treated by foreign doctors who had responded to the tragedy. No one would tell him how long he had remained buried under the rubble. He regained consciousness in hospital, with severe injuries, twelve days later but remembered nothing of what had happened. Martin was also told that his father had been killed in the earthquake, an extra tragedy in this young man's life since losing his mother just two weeks after he was born.

Rural mountains

The Trol Chouchow district covers a small range of mountain terrain rising west of Petit Goâve and stretching west for about 12 miles toward Miragoane, bulging upward between a southern boundary of National Route 2 and the sea to the north. Access is gained by a dirt track that leads north off the main highway and climbs for several miles up through beautiful wood-and scrub-land dotted with communities. Durisy is high up in the mountains that rise up steeply to the south of the coastal city of Petit Goâve. A track, part concrete, part dirt and bare rock, winds its way upwards and along the ridges of the mountains, through quite lush woodland and undergrowth on the partially cultivated slopes. After several miles, this track brought us to the district of Durisy. At the centre of the village is a large, deep lake, stocked with fish. There is also a Government school and a small medical centre. Scattered around the district are many houses, often hidden by trees, or by the maize and other crops, such as manioc, peas and beans. Many of these homes are built on ledges cut into the steep slopes. Some are almost hidden within copses of mango trees, overlooking parcels of cultivated land. Other homes are in areas devoid of vegetation, on exposed ledges on slopes prone to slumping. It is a beautiful area, often above the broken cloud level on a sunny day, hidden in it when it is raining. We conducted several interviews both at the medical clinic and in various homes.

Participants at Trol Chouchow recalled hearing noises, like a violent wind, thunder, an explosion or a boiling or growling noise coming from the east toward them. The earthquake tremors were felt strongly in this district, leaving some participants complaining of muscle aches and tremors. Participants here had experienced fairly regular earth tremors through previous years.

Some were at home

Virgil (a different person from the Léogâne Virgil!) was a local pastor and school director, living in the western end of Trol Chouchow. He had arranged the customary prayer service for the early afternoon. After the meeting, he had gone home and was taking a nap, lying on the floor with his 4-year-old daughter sleeping on his chest. He was asleep when the earthquake woke him. He saw the ceiling moving back and forth, and then the ground beneath actually seemed to give way. The floor of his house split, just enough for him to notice the difference in the levels either side of the split. A lot of damage occurred to his house, which had been a family house passed down through the generations. Effectively, it had to be rebuilt. Once the tremors had subsided, he and his wife and children joined other neighbours in a safe place away from the dangers of falling properties. Another of his daughters tried to run away and was injured on her head by a falling block, for which she required hospital care.

Some people were in church

In December 2012, on a scoping visit to Haiti I met with a group of pastors in the village of Cadet, Trol Chouchow. One of them told me that he lived locally, and when the earthquake struck, he was inside a church with eleven others. They heard people yelling outside and houses falling down, so they got scared and started running to get out. The church building, built of blocks with a metal roof, was damaged, but no one was hurt. Stephenson was one of around twenty-three men, in a men's group, meeting at the church. As they finished praying together, they heard a noise like thunder, and the church building began to shake. However, they did not run; instead, they raised their hands in the air and prayed, "Almighty this is your will, what you want to do you shall do." Then they headed for the door. They saw a lump of concrete falling off one side of the building, but all got out safely. One man who had been more severely shocked by the event was taken to his home. The rest of the group then decided to go and check on other folk in the area and to advise people to find a safe place to stay, away from their homes, in view of continuing aftershocks and the threat of landslides.

Some people were out and about/visiting

At the age of 23, Cecile was still in secondary education, in her eleventh grade, a common education demographic for Haitians. She lived with her parents and brothers up in the mountains of Durisy, but during the week, she would go down to Petit Goâve to attend school. At the time of the earthquake, she was lodging there. That Tuesday afternoon, she was child-care for a 4-year-old boy, the son of a neighbour. Her room was on the second floor of a two-storey building. She briefly left the boy while she went to take a shower and then returned to watch television with him. When the earthquake struck, the house collapsed, but not before she managed to escape, along with the neighbour and boy. In the brief moments of the earthquake, they had also witnessed other buildings collapsing and seeing many people dying. She and the neighbour went on a search around the local hospitals, looking for people they knew had survived. What they actually saw was horrific: people who were dead, who had lost body parts, one woman whose belly had been torn in two with her intestines hanging out. All the while as Cecile was having to witness these sights she was worrying about what may have happened to her own family back in Durisy. In the immediate days after the earthquake, her lodgings having been destroyed, Cecile joined most other survivors in a tented encampment, in open space outside the city.

In this chapter we have endeavoured to convey from Haitian narratives, what it was like physically to experience one of most the catastrophic earthquakes as far as human loss is concerned. However, to further the reader's ability to be affected by the event we must come next to the psychological and emotional impact on survivors.

Notes

1 Sobrino, *Where Is God?* 12.
2 Sobrino, *Where Is God?* 12.
3 The Haitian Creole, *blan*, is a term for any foreigner.
4 We interviewed 160 survivors, conducting ethnographic in-depth interviewing, field journaling and observation. Participants were selected using convenience sampling of those who were available for interview, some of whom would then refer us to people they knew, adding, in a snow-ball effect, to our data sample.

See Zoë Bennett, et al., *Invitation to Research in Practical Theology* (Abingdon: Routledge, 2018); A. Bryman, E. Bell, and J. Teevan, *Social Research Methods*, 3rd Canadian ed. (Don Mills, ON: Oxford University Press, 2012).

5 Both the bays of Grand Goâve, where three residents of Grand Goâve drowned, and Jacmel, where the UN base in Jacmel flooded, experienced a tsunami. See Herman M. Fritz, et al., "Twin Tsunamis Triggered by the 12 January Haiti Earthquake," *Pure and Applied Geophysics* 170 (2013). doi: 10.1007/s00024-012-0479-3; Matthew Hornbach, et al., "High Tsunami Frequency as a Result of Combined Strike-Slip Faulting and Coastal Landslides," *Nature Geoscience* 3 (2010). doi: 10.1038/NGEO975. Some 3,500 people have been killed by tsunamis in the Caribbean since the mid-nineteenth century, according to the National Oceanic and Atmospheric Administration, NOAA, "Full-Scale Test Today of Caribbean Tsunami Warning System," *UN Educational, Scientific and Cultural Organization*, March 13, 2013.

6 Wilentz, *Farewell Fred Voodoo*, 213–26.

7 William H. Bakun, Claudia H. Flores, and Uri S. ten Brink, "Significant Earthquakes on the Enriquillo Fault System, Hispaniola, 1500–2010: Implications for Seismic Hazard," *Bulletin of the Seismological Society of America* 102, no. 1 (2012): 18–30. doi: 10.1785/0120110077.

8 December–February are the months of lowest average rainfall in Haiti. The main hurricane season runs from May through November, with the most rainfall during May and September–October.

9 Prior to the earthquake, around 41% of Haitians were formally unemployed. According to USAID definitions, the labour force rate is the proportion of the population aged 15 and older that is economically active. Figures from USAID Economic and Social Database. *Haiti Has Both Formal and Very Informal Forms of Employment*. Online: https://eads.usaid.gov/esdb/data/country/profile.cfm?profile_id=9&country_id=332×pan=10&filetype=1. Accessed: 05/03/2015.

10 People resident in the U.K. may be interested/surprised to learn that there are between twenty and thirty earthquakes that are felt by people in the U.K. each year, with hundreds of smaller ones recorded by sensitive instruments (see British Geological Survey, www.bgs.ac.uk/discoveringGeology/hazards/earthquakes/UK.html).

11 Trauma psychologists Başoğlu and Şalcıoğlu recommend that people living in seismically active zones should have the opportunity to experience the sensations of an earthquake in an earthquake simulator, to acclimatise themselves to the sensations as a part of their control-focused behavioural treatment plan. Their view is that being acclimatised can reduce the risk of being paralysed by fear when the real thing happens. See Metin Başoğlu and Ebru Şalcıoğlu, *A Mental Healthcare Model for Mass Trauma Survivors: Control-Focused Behavioural Treatment of Earthquake, War, and Torture Trauma* (Cambridge: Cambridge University Press, 2011), 103–8.

12 As Gina Ulysse says, "Goudougoudou Doesn't Sound Nearly as Terrifying as the Experiences That Most Folks Will Recount When You Ask Them Where They Were That Afternoon When It Happened" (Ulysse, *Why Haiti Needs New Narratives*, 25).

13 These narratives accord with the scientific analysis. See R. Douilly, et al. "Three-Dimensional Dynamic Rupture Simulations: The M_w 7.0, 2010, Haiti Earthquake," *Journal of Geophysical Research: Solid Earth* 120 (2015): 1108–28. doi: 10.1002/2014JB011595.

14 Sven-Ake Christianson and Elizabeth F. Loftus, "Memory of Traumatic Events," *Applied Cognitive Psychology* 1 (1987): 225–6.

15 Interviewees came from different districts of the capital city, of which Port-au-Prince is one district, at the centre of the metropolitan. We are using Port-au-Prince in this chapter to refer to the metropolitan district.

16 *Machanns* are the market and street vendors who make up an important part of the economic and commercial life of Haiti. The majority are women.

2 When we were shocked

It's something I hope I never live through again. It's a thing that . . . it was very hard for us and it still lives in us. . . . It's a thing that kind of lives in our psyche, something that continues to haunt us, to live within us, the remembering of that.

(A Haitian survivor)

Introduction

Composing the previous chapter was not easy; this one has been even harder. We try to convey the psychological, emotional and spiritual effects from what survivors heard, saw and smelled as we recorded in the previous chapter. Haitians are well used to witnessing horrible things, sometimes routinely; this only emphasises how much worse were the effects of the earthquake, an event they had never before experienced.

Our approach is first to reflect generally on aspects of physical and psychological health that pre-existed as well as those that came after survivors' experiences of the earthquake. We then focus specifically on the predominant factor to which the narratives of our participants drew attention, namely their psychological encounters with the dying and the dead as significant aspects of their own shock and trauma.[1]

Of course, earthquakes never strike people living within a psychosocial vacuum. The Haitians we interviewed were living with a habitus of ailments and stresses, on top of which their earthquake experience just dumped a whole lot more stuff for them to bear and to adapt to. Furthermore, by the time we interviewed our participants, three years after the earthquake, they had also experienced a shocking outbreak of cholera, only adding to their burden of suffering and fear. As Hauerwas has said, we need to be able to "negotiate existence without illusion or deception." Pre-existing and succeeding health issues therefore need to be taken into account when trying to understand an earthquake in the Sobrino and Hauerwasian senses.

Encouragingly, in our survey of 215 respondents, 96% indicated they had no pre-existing psychiatric/psychological illness for which they had been treated. The highest proportion recording prior psychiatric illness came from Cité Soleil, where stress in normal life can be a dominant factor. Even so, respondents there, and overall, indicated very low levels of post-earthquake suicidal tendencies. Psychological issues therefore did not seem to form a significant background factor overall to the earthquake experience for our non-random, statistically unrepresentative cohort.[2] Theodore, the Director of Psycho-Social Services and Mental Health at the *Hôpital Universitaire de Mirebalais*, informed us that for Haiti's population of over 10 million there were just ten active psychiatrists.

From the narrative data from our interviews, we found a range of psychosomatic sequelae, including post-traumatic stress (*not* disorder) related anxiety, hyper-vigilance, nightmares/dreams and flashbacks, which would often be inter-related with somatic sleep-disturbance, headaches, palpitations, malaise and raised blood pressure. The majority of symptoms came from urban areas, where the destruction and carnage were greater and more publicly visible. In Petit Goâve, for example, Micheline found that relatives would often confuse her with her deceased cousin. This confusion was due to those relatives, having witnessed the damaged body of the deceased young woman, not wanting to accept the reality of the death. Relatives said they preferred to believe that she was still alive and would return some day. Thus, whenever they saw Micheline, who looked very much like her dead cousin, relatives would think it was her cousin. Also, Micheline's aunt found it hard to cope when visiting her mother, because when her aunt saw Micheline combing her mother's hair, the aunt would cry and ask who would take care of her now that her first-born daughter was dead.

Dreams, nightmares and flashbacks to scenes of death and destruction were common among our participants. These phenomena could last up to six months or more for some, but for most, they lasted a much shorter period of time.

After the earthquake, there was a flood of Western mental health professionals into Haiti, but given the time frame within which most arrived and departed, they could give little assistance above providing basic psychological first-aid. Theodore and Felicie, his Haitian psychologist colleague, set up psycho-social programmes, using mobile clinics to provide psycho-social care in the Internally Displaced Persons (IDP) camps.

A notable socio-healthcare consequence of the earthquake was the social and economic deprivation caused by the destruction of services and infrastructure, which led to women in the IDPs discovering their own bodies as sources of income. This was described to us by a social worker at the State University Hospital of Haiti:

> Even parents put their kids into prostitution so they could get money and it happens that the child get rich. And because the houses were

> destroyed they didn't have anywhere else to go, they didn't have money, they were in a tent living with no restrictions with other people.

When it came to physical health, pre-existing issues were numerous. There were dermatological, respiratory and diet-related problems. There was a host of viral/bacterial/parasitic diseases (malaria, dengue, typhoid, tuberculosis). Problems with hypertension and cardiac-related issues were relatively common, as were issues around arthritis and rheumatism. We also learned of sexual and maternal health-related problems.[3] On top of these issues came those produced by the earthquake experience. Broadly speaking, these were of a musculoskeletal nature, sleep related, dietary issues and water-borne diseases.

Musculoskeletal injuries

Musculoskeletal injuries were prominent. A common symptom in the Trol Chouchow mountain district was shaking in their limbs. Local doctors put this phenomenon down to stress following their earthquake experience.

A Haitian proverb says, "A foreseen disaster should not kill the handicapped" (Creole, *Malè avèti pa touye kokobe*). Unfortunately, this one did kill and maim. Among the 2,000 "refugees" from Port-au-Prince who turned up at the Mahanaim IDP camp, at Archaie, 27 miles north of Port-au-Prince, were many earthquake amputees as well as those psychologically affected. Assisting amputees gain access to prostheses was a part of the programme implemented by Felicie, the psychologist working out of *Hôpital Universitaire de Mirebalais*. It was Joseph's experience of witnessing amputees around Port-au-Prince prior to the earthquake, attempting to cross the traffic-clogged streets, that led him to have an immense burden to address the huge increase in the problems for amputees after the earthquake. It inspired him to develop a project he had been dwelling on previously, namely to help amputees access prostheses. Around 4,000 amputations were recorded immediately following the 2010 earthquake, but higher estimates have also been suggested.[4]

The policy of precipitate amputation, or "war/guerrilla medicine" immediately following the earthquake, reveals something of the enormous pressure placed on responding medical services in a country that struggles with basic medical infrastructure and where this kind of disaster can lead to potentially fatal infections from injury.[5] Even so, in Haiti the disability created by an amputation, especially if there is no subsequent access to prostheses, is more than likely cruelly life changing for the patient and any dependents. Dr. Paul Farmer, whilst recognising these social stigmas for amputees, reflected upon the clinical and social dilemmas medics faced when treating cases of serious crush injuries caused by the earthquake.[6]

Chancy, on the other hand, questions the medical and ethical case upon which mass amputations were carried out.[7]

Sleep-related problems

Some participants spoke of sleep-related problems, especially while living and sleeping in the open air and in the temporary accommodation of the "tent cities." In the immediate aftermath of the earthquake, many survivors were kept awake with anxiety caused by subsequent aftershocks. Armand told us that it was hard to sleep during the night whilst you were concerned for the protection of your children. Alex found his sleep affected by news items about the earthquake and media items drawing comparisons between the earthquake in Haiti and those in other parts of the world. Alexandre recalled his sleep being affected for a month after the earthquake. Agata, whose brother was fatally injured in the earthquake, found her sleep disturbed by recurring dreams of her brother. Apolline said his sleep sessions never last long. He may close his eyes to sleep but within a quarter of an hour, he would be awake again. He said he was desperate for sleep. Just thinking about the earthquake during the interview made him depressed. Cecile had bad dreams and flashbacks for more than five months. However, for most, their symptoms relating to sleep subsided within a few months of the earthquake.

Dietary problems

Diet-related issues, coupled with a general loss of appetite, could also be a problem through lack of access to good food or from preoccupation with the care of family or injured relatives. A number of people spoke of having little or no appetite for eating, mainly in the immediate aftermath. This was because they had seen dead bodies lying in the streets or trapped in collapsed buildings. Augustin said the shock of seeing so many people die, together with all the structural damage, deprived him of his appetite altogether for around five to six months. The stench and sight of the decomposing bodies deterred others from eating. Alberto, after seeing so many dead bodies piled up in the streets said, "Even if you were hungry you didn't feel like eating anything." Two participants told us it was the urgent search to locate missing loved ones that also affected people's appetites.

Water-borne diseases

Arising indirectly from the earthquake were water-borne diseases. One participant spoke of catching typhoid fever, possibly from an infected water source, while living in a temporary camp in the Léogâne area. Some spoke of their own, or of others', direct experience with cholera: a particularly virulent strain of the bacteria *Vibrio Cholerae* was imported from Nepal, where it is endemic, via some UN military staff ten months after the earthquake. More immediately, the geological impact of the earthquake disrupted natural water sources and courses, as well as damaging existing infrastructure servicing water and sanitation.

Farmer is right to remind us, "Health care does not exist in a separate universe from politics. Fiscal policy, infrastructure, wages, taxation – all affect the practice of medicine."[8]

In responding to all of the above physical maladies the Haitian health system proved totally unprepared and ineffective. This is not to suggest that medical staff were inept or negligent in their response and care. After all, they too were survivors. Many healthcare staff died or were severely injured in the earthquake. Those who were able rallied to their hospitals, or to their makeshift equivalents, where they did their best with the limited equipment and supplies available. The earthquake had been unannounced and sudden, and there was no preparation, no stockpiling for it and no rehearsals. The medical system was ordinarily broken, one where patients have to pay for their care and staff function on poor and sporadically delivered wages. Staff needed security at night, and there was none available. Ironically, within days, the system became reinforced, if not overwhelmed, by the international response of medical expertise from around the globe. Overnight, Haiti became medically equipped like never before!

This brings us to our main focus.

Psycho-spiritual experience with death

Figures for the earthquake dead have varied from as low as 85,000 to as high as well over 300,000. One study reported the "death toll of the Haiti earthquake will never be known, but the value of between 122,000 and 167,000 seems the most reasonable estimate, with a preferred value lying somewhere around 137,000 deaths."[9] Another study estimated that 158,679 people in Port-au-Prince alone died during the quake or in the six-week period afterwards owing to injuries or illness.[10]

Around half of our participants spoke explicitly of some kind of experience with death connected to the earthquake. Of those we surveyed, 71% had experienced the death of a close friend or loved one from the earthquake, with this figure rising as high as 81% in Cité Soleil.

To enable us to grasp the variety of experiences with death we narrate these in a number of specific contexts in which our participants encountered death.

Child deaths

To bring home the significance of child deaths, Dominique described the experience of a young friend of hers:

> Because you know someone may have only one child, and the child disappears. There are those that came, become mentally ill, because of it. I know someone who had a baby, her only child; her baby was trapped under rubbles. They did all they could to get the baby out, but they couldn't. The moment the mother came to what was left of her house, and having heard where her baby was, she immediately became mentally ill.

The impact of seeing their own twin sons run away from their father and die left deep emotional scars on Jonaissant and his Dominique; all the more so, as the boys both wrenched themselves out of Jonaissant's grasp, as he clung onto the hand of the young daughter.

Thirty-six percent of Haiti's population is aged 0–14 years, and 45% is under 18. The Haitian earthquake of 2010 took a dreadful toll on children. The visibility of this for many of our participants was even more shocking.

For Claire, the loss of her baby grandson, who died beneath the rubble of their collapsed home in the Carrefour district, was heart breaking, both for her and the baby's young mother and for Claire's younger daughter Grace, the baby's doting aunt. In the case of Raphael, whose house collapsed on him and killed his niece's husband and their baby, what hurt him most was the death of his teenage son and heir and the fact that they could not give the boy a normal funeral due to the lack of funeral services providing embalming and coffins. Instead, he rushed him, along with the children of neighbours who had also been killed, to a private, family cemetery.

Adele knew many people who died in the earthquake, but it was mainly the sight of so many boys and girls lying dead that upset her, making her wonder what she would feel like if this happened to her own children. Apolline described to us the appalling fatal injuries suffered by a baby she saw when a local church building collapsed onto rooms at the back of the house she rented.

The death toll in the schools presented a particularly tragic sight. Desiree, who had gone out of the school where she was working just before the earthquake struck, witnessed the school collapsing and said how shocking it was for her to return and learn that the majority of those inside had died.

Marguerite recalled that she had not gone to school until later in the day on that January 12th. She was in a second floor classroom when the building collapsed. Though badly injured, she survived but saw a student killed by a block falling on his head and a teacher killed while escaping from the building. Afterwards, Marguerite remembers parents coming to get their children from a school in which thirteen children had died.

When we interviewed Emile in 2013 in a school in Petit Goâve he told us he would rather die now than go through another terrifying experience like the 2010 earthquake. The event had caused great fear among the children. For him, personally, the worst thing had been hearing the cries for help from children trapped under the rubble of their school and his being unable to do anything to help them.

When we were interviewing folk in the Trol Chouchow mountain region, some parents spoke of their great fear at the time of the earthquake because their children were away in schools in Petit Goâve. Many children in Petit Goâve schools, who had come from outlying areas, were killed. The Trol Chouchow children had witnessed these deaths and then spent the night sleeping beside the main road before making their way on foot the next day the 11-mile trek back home. A number of these children did not want to return to schools in Petit Goâve, such was their level of fear from what they had witnessed there.

No chance to say "Goodbye"

For some people, the most distressing factor came from them not seeing their loved ones alive again after they had parted earlier that day. Others spoke of relatives whose bodies were never recovered, or, if they were, of the bodies having been taken off without notice to the national mass grave at Titanyen. A woman we met in Petit Goâve told us about her husband, a police officer, leaving their home in Carrefour early that fateful day, to go and work in Delmas. While at work, he was struck and killed by falling masonry. Well-meaning friends did not tell the wife of her husband's fate until some days afterwards for fear that telling her might harm her pregnancy. While visiting his orphanage, Manuel explained to us that the two young girls who had just introduced themselves to us last saw their parents going out from their home shopping on the day of the earthquake, but they never saw them again.

The *machanns* who operated in the shade of the galleries of major buildings, or who set up their stalls adjacent to large buildings in the city centres, were very vulnerable to falling masonry from these collapsing structures. Angelique had a cousin in this position, and they have never been able to find her body.

By the time he had arrived home after the earthquake struck, Antoine discovered that his 12-year-old son had been killed as their house had collapsed upon him. The body was never recovered. What helped the father and mother resign themselves to their loss was seeing so many other children's bodies lying along the route to the open space where people were gathering for the first night. Aristile lost his brother in Port-au-Prince and they never found his body. It broke Simon's heart to learn that his closest friend, who had that same morning so generously loaned him a large sum of money, was crushed beneath offices where he worked and was never seen again.

Bodies, everywhere!

For Patrick, seeing a dead body or two was nothing new. In traffic incidents, or even when you wake up in the morning, he said, you could find the body of somebody who had been killed, lying in the road or on the pavement. However, Patrick had never seen death on such a scale as he had from the earthquake.

Of those living in the cities and towns, a number of our participants spoke of having to make their way home from where they were when the earthquake struck, usually to check on family members. In making such journeys, some spoke of having to walk over the bodies of the dead and of the injured lying in the streets. Simon had experienced the same as he walked over 10 miles, via the National Palace in Port-au-Prince, to Carrefour. These sights were terribly upsetting and would linger in their dreams and memories for many months afterwards. On witnessing the devastation and death in his city of Léogâne, Simeon remarked, "People in front of you who [were] agonising, dying. You don't know what to do. There's no water, there's no drinking water. There's no food. You can't think about food, you can't think about anything." Metellus lamented that he too had never seen death on such a ghastly scale as he related to me in his broken English,

> Everything, ever since I think bad about the earthquake, make me feel, you know I saw, is only time I saw most of my people, of my nation, dead in the street, like pigs! Now it's like I remember the time I went back to the town in Léogâne on January thirteen . . . it was like sand – people was like sand on the beach, as a lot of people dead, lot of them. So many people, squashed and broken.

Seeing death and hearing dying

Far more participants said they saw people who were already dead than who actually witnessed the process of dying. However, some did witness the most appalling deaths of relatives, friends and neighbours. For example, Michel had one of his sons, a student at university, calling out to him for help from beneath the masonry of their collapsed house. However, they were unable to find the tools necessary for digging him out, so, by the following day, when they managed to get to him, he had died. Adelaide saw a house collapsing onto a 9- or 10-year-old girl as she tried to run out of the house and this had affected her deeply. Alexandre remembered seeing a man trapped between a wall and a crashed *tap-tap*. The injured man was crying out for help as lots of people tried to remove the *tap-tap*, but the man died of suffocation. Agata, a young woman who had come to the Cité Soleil medical centre, told us that she had seen a boy being crushed by a wall, next to her own house, which had collapsed on him. She had then gone to look for her brothers who had been playing football. She came across them carrying a brother who had injured his head, arm and side and who had a broken leg. They took him from one hospital to another but both hospitals were full. The severely injured brother died outside one of the hospitals, in Agata's arms, with her mother in inconsolable grief, wailing as she paced up and down beside her.

Andre witnessed a primary school collapsing on the students present there, resulting in many deaths. With tears streaming down his face as he just walked around the streets in a stupor, he had to sit down, so overwhelmed was he by the sight of so many dead bodies. What distressed Angelique also, as she walked the streets of the capital, was the number of mothers and fathers crying out because they had lost children. On his walk back to Petionville, where he and his family were living at the time, Alberto said his heart was broken as he witnessed so many people lying dead in the streets.

Joseph reflected that January 12th, 2010, was the first day when he witnessed a whole bunch of people dying on the spot. He went on to describe the chaos he discovered when he came out of his college in a district of Port-au-Prince:

> Yes, there were cars that were colliding with one another. There were those that were getting a transportation car, a vehicle, and they died. And the scene behind me, there was houses falling down. There was this little girl who had just got in from school, and a house or business collapsed on her. And me, I almost got hit by a car while running to cross a street. And when I did cross the street, on the other side of the street I saw that houses had fallen and there were people trapped under them.

Marie was traumatised by witnessing a close friend and their child being killed as their house collapsed next door to hers. Helene, only 15 years old at the time, was one of the few we interviewed who still exhibited the signs of post-traumatic stress three years after the earthquake struck, killing her mother as they both attempted to escape from their rented house in Petit Goâve. She had actually seen her mother crushed by a wall falling on her.

Worshipping and dying

In some cases, people our participants knew had been gathered in church services when the earthquake struck. This was the case with the husband of Aimee. She had been unwell on that day, but her husband wanted to attend the service at the local church, so she stayed at home while he attended the service. The church building had collapsed and he had been seriously injured. He died as others tried to get him to hospital.

Augustin recalled losing a cousin who had been very close to him. He had been in a church service at the time and twelve people had died there. A group of Christians had gathered with their pastor in a church member's house in Léogâne. They were holding hands in a circle to pray for a woman's sick son when the earthquake struck. Eight of those gathered died inside that building. Among the dead was Paul's wife, and when he learned of her death he said that he went crazy, losing his mind. He went walking for miles around the roads of the district in a totally disorientated state until a man found him and brought him back to Léogâne. The Catholic Archbishop of Port-au-Prince, the late Monsignor Joseph Serge Miot, had knelt in prayer on the balcony of his offices. When they collapsed in the earthquake, he plunged down to the ground beneath and was killed.

Praying and worshipping did not guarantee immunity to violent death for these Christians in Haiti.

The agony of the injured dying

The earthquake produced a great many injured casualties who later died because of the failure of medical facilities and/or the lack of heavy lifting equipment to effect a more urgent rescue. In some cases people suffered enormous distress from seeing and hearing the cries and groans of the injured who died from hunger or thirst rather than from their injuries while trapped inside collapsed buildings. Apolline recalled:

> My problem with the earthquake is that a lot of people that were trapped under the rubbles did not die right away. Some of them lasted for fifteen, twelve days. Pain and hunger killed them. . . . Because, being stuck under something, and by the time for someone to help you, and there were no one to help you, you can't get out by yourself. The only thing possible is death.

With such thoughts, can you imagine how Apolline felt when her cousin's body was still fresh when recovered from under the rubble several days after the earthquake, indicating that she had lain there alive up until just prior to being found?

In Léogâne, Andre was left profoundly upset after seeing a woman, trapped head first between two concrete floors that had collapsed on her. She was alive, so some people tried to lift the concrete floor to free her, but the tractor they used could not help, so she died there. The remarkable survivor Estelle wept as she recalled some of those trapped alongside her slowly dying under the rubble of the Hotel Montana. Unbeknown to her at the time eighty other guests, staff and family members also died under the terrible collapse. For Marcos, witnessing the injured dying was more than he could bear, especially the little girl with the skull split open and brain exposed who was carried to him by a traumatised mother. Though Marcos desperately sought medical help for the child, by the time he came back to her a paramedic friend informed him the girl had died and the mother had just walked away into the crowd.

The smell of death

One of the features of mass deaths in a tropical climate is the lingering stench of death. This was certainly the case for those living in major population centres in Haiti affected by the earthquake, and it just accentuated the sense of horror from what had happened. Cité Soleil participants spoke of the terrible smell of death they encountered, especially in the city centre, where there were many bodies buried under collapsed buildings. Alexandre said, "And gradually the smell got worse and worse and it just felt as if there were more people who died than there were who were living." In fact, when the cholera struck, in October 2010, he said people initially believed it had been caused by the smell arising from decomposing bodies, a myth that is commonly invoked after a disaster involving mass fatalities and that often provides authorities, unnecessarily, with the notion for quick, mass burials.[11]

In fact, it was the smell of death that was a key factor contributing to the mass migration into the provinces that followed the earthquake. Claire said that she and her daughters had to move out of Port-au-Prince and go to Jacmel, in the south, because by February 2010 the smell of death was becoming so unbearable in the capital. Such foul odours were the norm for other centres of population as well, like Léogâne, Grand Goâve and Petit Goâve, where many deaths had occurred and where bodies had been left in the open. Sometimes this stench was coupled with personal tragedy for our participants. In the case of Marceline, she was in her concrete house, in the Léogâne district, along with her sister and partner, when the ground began shaking. She heard someone shouting, "There's a house about to fall down!" whereupon she ran out of her own house just as it collapsed. Immediately after the dust settled she asked where her sister was and no one could tell her. Three days later, their dog began smelling a bad odour around the ruins of the house. When they dug down, they found the remains of her sister and her partner, buried under the block walls that had fallen inwards onto them.

Dread in the mountains

In the mountainous regions of Durisy and Trol Chouchow, storms and earth tremors bring a terror of their own. Here the main danger does not come so much from collapsing buildings, but more from rock falls, slumping hill-sides and the opening up of ravines. Slopes can become liquefied as the water table changes and landslides carry away people as well as essential crops and cattle. In the mountain communities where we conducted inter-views, there was much less witnessing of death from the earthquake. Such deaths as occurred were usually the result of landslides or of houses of block or rocks collapsing inwards. At least two participants in Durisy pointed out to me on the mountain slopes above, a location where two people had been standing on the side of a hill, one inside his house. When the slope gave way it buried one of them. A girl had been up a tree searching for breadfruit while a friend was waiting below to catch the fruit as it fell. One fruit fell down into a small ravine and the friend clambered down to recover it. When the earthquake happened, the ravine walls collapsed on her.

Disposal of the dead

Arguably, no nation would have the capacity to cope with the number of deaths in Haiti at the time of the earthquake, let alone the degree of destruction to hospitals and to mortuary facilities and to their staff. The severely damaged State University General Hospital in Port-au-Prince registered 2,200 human remains between 8 a.m. and 3 p.m. on the day after the earthquake. Gupta and Sadiq interviewed Pierre Yves Jovin, the morgue director at the hospi-tal, who said that the hospitals receiving the dead became overwhelmed by January 13th. Around 5 p.m. that day, the Government passed a rule stating that all bodies had to be collected by the Central National Equipment (CNE) and disposed of in the communal grave at Titanyen. Techniques used, Jovin reported, were archaic.[12] There was no indication of any tagging of bodies for future identification purposes, no pathology and no post-mortems. CNE is the Government public works agency, with heavy machinery for road con-struction. It had no prior mandate for, nor experience of body identification.

The only exceptions concerned identifying foreigners. Gupta and Sadiq reported, "No effort, whatsoever, was made by Government to identify the dead, except foreigners. The facilities for forensic identification, like DNA testing, dental records and fingerprinting were not available."[13]

In Haiti, it seems that virtually every aspect of the accepted international protocol for mass-fatality management (MFM) was disregarded, for whatever reasons. Notable exceptions to this, however, relate to the way the bodies of some foreigners (U.N. related) were dealt with. For example, Gupta and McEntire et al. record that the U.N. contracted with Kenyon International Emergency Services, a company specialising in mass deaths, to locate and handle deceased U.N. personnel.[14,15]

Haitian anthropologist Gina Ulysse maintains, "The state treats the dead as it does the living," and nothing demonstrated this so graphically as the State edicts ordering the bodies of Haitians to be collected by lorries and dumped in mass graves. Ulysse sees this as symbolic of the Haitian State abandoning the nation.

During Roger's visits to Haiti in 2010 and 2011, he heard of the national mass grave, located in some treeless hills just inland off National Route 1, north of the capital. It was a sombre experience to visit this enormous expanse of gently sloping hillside of beige coloured sand and gravel, punctuated with so many rows of small, symbolic, black wooden crucifixes. At the rear was an enormous mound rising up out of the slope with larger white

Figure 2.1 Hundreds of bodies collected in a parking lot at the main hospital are dumped into a truck in Port-au-Prince, two days after the earthquake hit the city late Tuesday afternoon.
(January 14, 2010)

Source: Photo – Timothy Fadek/Redux/eyevine

crucifixes lining a wide strip of purple carpet and an even larger crucifix standing atop the mound.

Being a culture saturated in Vodou, the significance of ancestors is accorded high spiritual recognition, requiring careful ritual veneration. The matter of body disposal and funeral rites seemed to us an appropriate aspect for exploration during the interview process because of possible psycho-spiritual effects. Researchers Gupta and Sadiq comment:

> Cadavers therefore need to be treated in a way befitting the ritualistic and religious practices; otherwise, it may leave a Zeigarnik effect . . . on the surviving family members and the community. For instance if an unidentified body is cremated or buried without markings, it may foreclose the possibility of identification forever. . . . The way dead bodies are dealt with reflects how the living are treated and respected.[16]

This could mean psychological, and even physical, distress being incurred by surviving relatives. According to Dr. Paul Farmer, who was in a meeting with former U.S. President Bill Clinton in New York on January 13th, 2010, to discuss the response to the earthquake, Clinton specifically requested that "something be done to preserve the bodies of the Haitian dead," because "[t]hese people deserve a chance to bury their dead." Our interviews bore out the fact that reality on the ground in Haiti made it clear that as time went on the chances of Clinton's wishes being upheld receded fast.

After the earthquake, many bodies were recovered by Haitian citizens and left on the streets. Many of those trying to recover bodies found their way to those bodies by following their noses. Clea recalled seeing many bodies being collected and taken to the local mayor's premises. She was so distressed by this that she could not watch for long. Unless someone came to remove them first, other bodies were disposed of by family or neighbours, simply dousing corpses with petrol and burning them in situ. Elizabeth witnessed this happening routinely to the piles of corpses left on the streets in Port-au-Prince. It also happened with the sister of Marceline, in Léogâne, after which she took the ashes and buried them. Marcos related to us how, in Grand Goâve, with the smell from a pile of dead bodies becoming unbearable, the authorities threw petrol over them and burned them.

We had heard that this policy of using the mass grave arose not just from the mistaken fears of disease being spread but also from seeing roaming, hungry animals devouring bodies. Furthermore, the hospital and funeral directors' morgues became overwhelmed and could not cope with storing the dead. In fact, people recalled seeing bodies lying on one side of the street while on the opposite side survivors were sleeping, and feral dogs were roaming in between. There was also the factor of a lack of electricity supply for the necessary cold storage of the dead. Ignace was shocked by the piles of bodies that were left outside funeral parlours in the capital, awaiting disposal in some way. A mechanic working in the St Croix hospital, Léogâne, told us that all the bodies that were brought to that hospital were removed to the mass grave because there was no electricity supplying the hospital morgue. In Petit Goâve, so Heloise informed us, a cousin who died from her injuries had been brought to the local morgue for burial, but such was the number of bodies and the absence of electricity, that all had to be removed to a mass grave.

Reactions from our participants to realising that their loved ones were disposed of in a mass grave were mixed. Some regretted it deeply, but others seemed to have little problem. This did not mean, though, that there was great approval of the mass grave policy. When we asked Paul, in Léogâne, how he felt about seeing his wife taken off his hands for disposal at a mass grave under instructions from the local mayor, he replied: "I was not happy! She was a school teacher and I spent twenty-four years living with her, and then to later see, to witness that's how her body's being disposed of . . ." In Cité Soleil, Rene, as an only child, found the discovery of his disabled mother, crushed beneath the rubble of the house where a friend had been nursing her, was unbearable and all the more so as she was disposed of, one week later, in the mass grave.

Figure 2.2 View of the mass grave site in November 2010, Titanyen, with the Bay of Port-au-Prince in the background.

Source: Photo – Roger P. Abbott

Some people did all they could to recover their deceased and to dispose of them using their own private funeral rites and graves, thus giving themselves some semblance of satisfaction over the disposal of their loved one(s). Martin expressed how comforting it had been for him to take his dead father, killed in his house collapse in the Delmas district, back to the Ile de la Gonāve for a funeral on January 23rd, 2010. Tragically, for Emilie, it took five months before they could extract the bodies of her mother and sister from beneath their collapsed home in Port-au-Prince. This was because those who were paying for the three-storey block to be broken up, so the bodies could be recovered, would sometimes lose hope and give up, at which point the work would stop. When Emilie was eventually able to recover what was left of her mother and sister, she held a private symbolic funeral. Leroi, who ran several funeral homes, placed his 4-year-old twin daughter on a board, with the clothes she had on when she died, and laid her in the family tomb. There was no funeral nor could they preserve her body for interment, but there was no way he was going to allow her to be taken to a mass grave, even though most of Leroi's funeral parlours had been destroyed in the earthquake. Michel would not allow the body of his daughter to be taken to the mass grave like so many others. Instead, he carried her to the local cemetery for interment.

On the other hand, some participants seemed to understand that the state of emergency required mass body disposal of a less dignified kind. Yve found his sister-in-law decomposed under the rubble of a house eight days after the earthquake, and such was the state of decomposition he felt they had no other choice than to have her removed to the mass grave. Wilmer had been a worker at the Lafito flourmill, near Titanyen, when the earthquake struck, coinciding with a huge explosion of propane gas cylinders in the mill. While he managed to escape with serious burns, the dead who were recovered from the burned factory were taken to the nearby mass grave. Wilmer said he thought that there was no way their deaths could have been investigated, and it was, therefore, understandable that they should go to the mass grave. Virgil, a pastor and teacher in the mountain district of Trol Chouchow, spoke of the great sadness he still felt over his brother being killed in the earthquake and then being removed to the Titanyen mass grave by the Government. However, he said, "But mentally speaking, intellectually, we know that God says man is dust and to dust he will return. But what hurts me the most is the fact that he [his brother] is not here to be able to watch his son grow." For Claire, her grandson was recovered from beneath the rubble nine days after the earthquake and was removed immediately to the mass grave.

Andrena Pierre et al. concluded from their literature review that since many people did not have the opportunity to find and bury their lost loved ones or see them buried in a mass grave, then "there may be an increase in ambiguity and uncertainty over the fate of the dead, with the risk of nightmares, worries and moral concerns when thinking about the dead."[17] We, however, found no indications of these concerns among our participants; we believe this had a great deal to do with the impact of their Christian beliefs mitigating the impact of traditional Vodou beliefs regarding the dead.

Conclusion

Our participants recalled truly horrific effects of their earthquake experiences, with some experiencing the terrors of being trapped beneath collapsed structures themselves, incurring physical injuries from which most have now fully recovered; others were spectators of similar things. For a few, the deep mental scars that were inflicted remain quite raw, while for others their memories are only retrieved by particular incidents occurring or when, years later, they need to recall those dreadful events (such as during our interviews). However, even in a culture where death is no stranger in public life, the experience with death presented by the earthquake was unique and unprepared for, and even more terrifying for that reason.

This taxonomy of the experience of death we have presented, though somewhat censured, is the most emotionally disturbing feature of our research. However, it is deliberately designed to convey the strength of horror those confronted with a natural disaster of these proportions go through and the understandable degree and range of psychological impact that experience with death, especially mass deaths, can cause.

Notes

1 "Shock" refers to immediate and temporary effects, whereas "trauma" remains, for many months or years (Shelly Rambo, *Spirit and Trauma: A Theology of Remaining* (Louisville, KY: Westminster John Knox, 2010)).
2 Data from formal medical records is unreliable in Haiti because many official records were lost in the earthquake.
3 Only one person mentioned AIDS, contra the prominence given to it by the international media. See Paul Farmer, *AIDS and Accusation: Haiti and the Geography of Blame* (London: University of California, 2006 ed.). A survey of illness six weeks after the earthquake found 44% of respondents had household members who had suffered diarrhoea, headaches and fever (Athena R. Kolbe, et al., "Mortality, Crime and Access to Basic Needs before and after the Haiti Earthquake: A Random Survey of Port-Au-Prince Households," *Medicine, Conflict and Survival* 26 (2010): 281–97).
4 A. Redmond, "A Qualitative and Quantitative Study of the Surgical and Rehabilitation Response to the Earthquake in Haiti, January 2010," *Prehospital and Disaster Medicine* 26, no. 6 (2011): 449–56. The Haitian Healing Hands of Haiti organisation in Port-au-Prince estimates 40,000 amputations after the

earthquake (Marie-Rose Chaperon, "Haiti: Limited Access to Healthcare Leads to an Amputation Epidemic," *Nova Southeastern University, College of Health Care Science*, December 12, 2013). See also Anahid Kulwicki, "Post Traumatic Stress Disorder (PTSD) in Post-Earthquake Haitian with Traumatic Amputations," *University of Massachusetts Boston, College of Nursing and Health Sciences* (November 2014): 1–35; Ronan L. J. Iezzoni, "Disability Legacy of the Haitian Earthquake," *Annals of Internal Medicine* 152 (2010): 812–14. doi: 10.7326/0003-4819-152-12-201006150-00234.

5 Myriam J. A. Chancy, "A Marshall Plan for a Haiti at Peace: To Continue or End the Legacy of the Revolution," in *Haiti and the Americas*, ed. Carla Calargé, et al. (Jackson, MS: University of Mississippi, 2013), 203, 211.

6 Paul Farmer, *Haiti after the Earthquake* (New York: Public Affairs, 2011), 113–15. For an example of difficulties facing amputees see "The Girl Rescued from Haiti, and Why I am Taking Her Back," *The Times Magazine*, September 20, 15: 19–22.

7 Chancy, "A Marshall Plan for a Haiti at Peace," 203–4.

8 Farmer, *Haiti after the Earthquake*, 23.

9 J. E. Daniell, B. Khazai, and F. Wenzel, "Uncovering the 2010 Haiti Earthquake Death Toll," *Natural Hazards Earth Systems Sciences* 1 (2013): 1929–30. The Haitian Government gave 320,000, but suspicions were cast over this being an inflated figure in the political interest of gaining maximum foreign disaster aid. Many use the 220,000 figure.

10 Athena R. Kolbe, et al., "Mortality, Crime and Access to Basic Needs before and after the Haiti Earthquake: A Random Survey of Port-Au-Prince Households," *Medicine, Conflict and Survival* 26 (2010): 281–97.

11 The Pan American Health Organisation and World Health Organization state that "dead bodies from natural disasters do not cause epidemics. This is because victims of natural disasters die from trauma, drowning or fire. They do not have epidemic causing diseases such as cholera, typhoid, malaria or plague when they die." And, "The risk from dead bodies after natural disasters is misunderstood by many professionals or the media. Even local or expatriate health workers are often misinformed and contribute to the spread of rumours." Pan American Health Organisation and World Health Organization. Online: www.paho.org/disasters/index.php?option=com_content&view=article&id=719&Itemid=931. Accessed: 11/12/2014.

12 Kailash Gupta and Abdul-Akeem Sadiq, "Responses to Mass-Fatalities in the Aftermath of 2010 Haiti Earthquake," Quick Response Research Report, University of North Texas; Natural Hazard Center, University of Colorado.

13 Gupta and Sadiq, "Responses to Mass-Fatalities," 8. They also report that Radio Frequency Identification (RFID) chip implants were used after Hurricane Katrina (2005). Gina Ulysse recalls being at a meeting at the centre for Latin American and Caribbean Studies in New York on January 20th 2010, when a Haitian survivor of the earthquake lamented that the initial rescue efforts by rescue agencies focussed on the rescue of white Americans/*blans*. (See Ulysse, *Why Haiti Needs New Narratives*, 9).

14 McEntire, et al., "Unidentified Bodies and MFM in Haiti," *International Journal of Mass Fatalities*, 30 (3) (2012): 318; Kailash Gupta, "Seeking Information, after the 2010 Haiti Earthquake: A Case Study in Mass-Fatality" (Ph.D. Dissertation Eric, 2013), 100. University of North Texas. Online: https://digital.library.unt.edu/ark:/67531/metadc271823/.

15 Ulysse, *Why Haiti Needs New Narratives*, 13.
16 Gupta and Sadiq, "Responses to Mass-Fatalities." The Zeigarnik effect, based on experiments by the Russian psychologist Bluma Zeigarnik (1927), means people are more likely to recall an uncompleted task than they are completed ones. Uncompleted tasks stimulate psychic tension in a person, which drives that person until they can complete the task (See Irving B. Weiner and W. Edward Craighead, *The Corsini Encyclopedia of Psychology*, vol. 4 (Hoboken, N.J.: John Wiley & Sons, 2010), 1873–1874).
17 Andrena Pierre, et al. "Culture and Mental Health in Haiti: A Literature Review," *Sante Mentale au Quebec* 35 no. 1 (2010): 33. doi: 10.7202/044797ar.

3 How our faith responded

> *The one, the one who made the shoulders; it is the same who distribute the crosses.*
>
> (*Haitian trauma therapy patient*)

Introduction

In the previous chapters, we have described survivors' experiences of the earthquake in terms of the physical and psychological impact upon them. This chapter presents the role religious faith played in the way survivors of the tragedy were affected and in their recovery process.

To begin with, however, why does religion warrant this kind of focus?

First, the religious demographic for Haiti is both straightforward and complex. Christians (Catholic and Protestants) dominated the Haiti religious demographic in 2010, forming 87% of the population; 11% were unaffiliated, and 2% professed folk religions.[1] The CIA Fact Book narrows the data down further, listing: Roman Catholic 54.7%, Protestant 28.5% (Baptist 15.4%, Pentecostal 7.9%, Adventist 3%, Methodist 1.5%, other 0.7%), Vodou 2.1%, other 4.6%, none 10.2% (2003 est.).[2] However, in view of the complex symbiosis of African folk religions and Roman Catholicism, called Haitian Vodou, little is definitive about these statistics in reality. Haitians may be 87% Christian, but they are also 100% Vodouisant, such is the cultural "biology" of Haiti.[3] The greater majority by far of our participants self-identified as Christian (Catholic or Protestant) and when questioned about any past or present associations with Vodou, nearly all distanced themselves from Vodou entirely or harked back to associations during their childhood in conjunction with their parents. We were assured by our Haitian research assistant, "You will never get a Haitian to admit to you that they are into Vodou." One *houngan* (Vodou Priest) and a *manmbo* (Vodue Priestess) we interviewed, however, confirmed to us that they had Christian clients who attended their services, including pastors, though they would not identify them by name.

Figure 3.1 Vodou in Haiti cannot be ignored: some of the accoutrements for Vodou ritual.

Source: Photo – Roger P. Abbott

A second reason for the religious focus is that traumatic events can impact religious belief, both negatively and positively.[4] Though the great majority of our participants found their faith a great support, this did not prevent them from experiencing severe challenges to their faith early on. Jonaissant's and Dominique's experiences clearly highlight their struggles in coming to terms with the emotional and cognitive dissonance they felt after the 2004 flooding in Gonaives followed by the 2010 earthquake. Jonaissant admitted,

> I feel desolate at times because I sometimes stop and think about how . . . through all the things I went through in Gonaives, in the 2004 flooding, and afterwards all the misery we went through in Port-au-Prince, and after the earthquake and how I lost my sons, and at times I feel like losing my mind, but God is . . . but God gives me, um, sanity . . . and I also took example of Job, who lost all his things yet God took care of him, and I know that God will also take care of me.

When Leroi and his wife lost their twin daughter while visiting his brother, it was emotionally devastating for them, especially for Leroi's wife, who had to be treated for PTSD in the U.S.A. Even so, they found it was their faith that got them through. When I asked Leroi, "You've spoken of your Catholic faith: how significant has that been for you through all these difficult times? Your bereavement, the loss of your daughter . . . and of everything you have lost?" His reply was,

> Well, that has been God. God has been my, my support, my backbone, my . . . being able to talk to you right now, not really feeling down: it's on account of God and the Virgin Mary. I have real faith in them and they help me very much, very, very, very. Anything I need, anything I want, I don't get it right away, but I will get it! And it comes so easily to me!

Alcee, from Cité Soleil, spoke of the challenge of the "mystery" the earthquake caused: "The mystery is, why would God let this happen, with all these people dying in the street, and the odours and smells coming from their corpses?"

Third, there was the apparent randomness of death during that time. Alcee related a session after the earthquake when he and a dozen Christians met to reflect on what had happened, to try and make sense of it and of God's role in it. They asked, "Can God just do this? Can God just destroy like this?" And what should they make of the fact that the Bible prophesied such events? After a whole day discussing these issues, they came to tentative conclusions mainly focussing on human factors, such as the Americans bombing Haiti and poor construction practices. Alcee concluded that God is impartial in his acts and judgements:

> God does what he needs to do or wants to do, but there will come a moment when God makes his judgment. What's really important, . . . what's really important for a person is for them to be thinking about their life, so that they're, whatever, whenever the moment comes, they're ready for it.

As if to illustrate the point, he said,

> Handsome young men and beautiful young girls who were just cast aside because of all this. What's important out of all this is not the body but the spirit and what happens to the spirit when you die. It makes no difference to God whether you've got a car or whether you haven't got a car; whether you've got beautiful clothes or whether you haven't got beautiful clothes; it's a leveller. So what it has shown us is that they in a sense were all the same, across the social bar . . . the social classes. In the university, there were doctors who were killed. To God we're all the same. We're essential. Doesn't make any difference

> whether you are wealthy and important or poor and unimportant, we're all on the same level.

Adolphie also wondered about his decision to leave the restaurant, to get money from one of his students, just three minutes before the place collapsed and killed everyone inside: "I – I left from the restaurant, like, three minutes, three minutes from there, who I am! I left there three minutes only, and all the people that I saw before I left, they all dead there!" The difference between living and dying could be down to a simple decision to stay a bit longer in school to swap jokes, Adolphie mused.

Fourth, we found that faith played a significant role in participants' interpretation of the earthquake therapeutically as well as theologically. Their theology provided a framework of understanding and explanation that brought them some cognitive and emotional peace in the wake of the initial terror and shock.

Religious themes

The main theological themes to which survivors appealed in their response to and recovery from their experience of the earthquake were the Bible and creation, but most significantly the doctrines of providence and hope.

Bible

For most participants the Bible was the main source and textual authority for their beliefs. Even illiterate participants had gained their beliefs from biblical information communicated to them by members of their family or church, or by Christian missionaries.[5]

Knowledge from the Bible had helped prepare people, albeit in hindsight, for when the earthquake happened. For many the Bible provided great comfort and reassurance that the terrible event was under divine control and purpose, even though most people had little understanding of what an earthquake was before it happened. Once they made a connection between the word *earthquake* in the Bible and the event itself, then this provided confirmation that the event, terrible though it was, was nevertheless under divine control. Xavier explained:

> With our faith, we knew about the earthquake; but since we are only human, only flesh and blood, we will always be scared; but we were warned that there will be earthquakes. . . . I did not receive any information concerning geology, but based on the Bible we knew that there would be disasters that will shake the earth.

Simon confessed that his faith had an enormous impact on him, and he could see from the Bible it was something that was prophesied. Because he saw that God's word was being accomplished this actually increased his faith. Jesus' prophetic words in Matthew 24:7 were an advance warning and form of education for both Augustin and Antoine.[6] Therefore, being aware of such biblical texts already, their faith was not affected adversely by the earthquake when it happened, even though the event itself was so shocking. Equipped with his biblical information, after the event Pastor Virgil and his wife became teachers on earthquakes to their neighbours, telling them, "Do you not remember, do you not know that even the Bible says that in the last days there'll be earthquake?" Heralding this message became their new role after the earthquake. In the case of Marie, she found that her reading of Psalm 46, where it said, in her words, "When he [God] strikes down his foot, the earth quakes," comforted her, that this was the work of God.

Creation

Whether amid an acutely dismal setting of creation in Cité Soleil or in the beauty of the mountains of rural Petit Goâve, participants were most affirming of a theology of creation. God was the creator of the natural world, that work of creation being achieved through the divinely spoken word and by the pre-incarnate Word, by divine fiat. The account in Genesis 1 was accepted as the pattern for God's work of creation, including for human life. Metellus, representing the view of many, said he had read in the Bible that God created all things in six days. In his case, he had also reflected upon what preceded God's existence or whether God is eternal. Angelique believed God first created the world and then decided to create humans, using Adam's rib to do so. Alex found that looking at the natural creation gave him an adequate explanation for where the world has come from, something he believed science could not do. For Charis, God created the world for humankind to enjoy. Student Andre had given serious thought to the matter of creation. Reasoning that the universe cannot create itself he came to believe in a creator God.

For some participants, their view of creation and of the "Fall" (Genesis 3) had a clear influence upon how they viewed the earthquake itself. During interviews, we usually asked what they thought caused the earthquake: for some their beliefs in creation were greatly influential. Those with a middle-class education, for instance, believed the earthquake to be a natural product of God's creating and sustaining work. Metellus told us he believed earthquakes were created by God in connection with the "plaques" (tectonic plates) designed to move alongside one another. He said he had learned this from people's experiences and from the internet.

In fact, a number of participants took the view that the earthquake was a naturally occurring event – a normal part of the way the physical world is designed by God to operate. Other, more religious purposes to natural hazard occurrence might also be included, but these would not negate the primary interpretation of the earthquake as a natural occurrence. A good example of this is the belief of Henri, a university student in Port-au-Prince. He believes earthquakes are natural events of creation but also events that, though not directly caused by Christ, still fall under the plan of Christ. He sees no conflict between these perspectives. Ignace, a young, educated, middle-class student, responding to our question, "Now you're a Christian . . . If somebody asked you why did the earthquake happen, what would you say?" replied,

> I'd tell them it is a natural phenomena [sic], and as a natural phenomena anything can happen. I would not disagree with people that says that it was the will of God because everything that happens is within God's control. And so, but I mostly tell them that it is a naturally occurring disaster, because just like cyclones, cyclones have their natural [way of working].

Gabriel represented a number of our interviewees when he indicated two reasons he believes the natural hazards of creation require a corresponding human responsibility. He commented:

> First and foremost, the earthquake is a natural phenomenon. It was all because here in Haiti we did not take the precautions to foresee such disasters. For me, that's my reason. My second reason I can say is that, in Haiti, we have three percent of vegetation. It is evident, having treated nature that way; it is evident that she would seek justice.

For many in the rural regions, who lacked significant education, their beliefs were more along the lines of the earthquake being a kind of awakening or judgement. Juliette's belief in God as creator had persuaded her that the purpose of the earthquake was to convince those who do not know him and those who sin too much that God has all the authority. Even so, Gabriella, as a result of scientific information supplied after the earthquake, believed the event could only have been a natural one.

Taking a Vodou perspective, Roberta, a Vodou *mambo*, with Catholic/Protestant sympathies, said she had three ideas about the earthquake: as a way of God touching Haiti socially, as a natural phenomenon and as an un-natural phenomenon. Of these three, she believed the first was the most significant reason. Her fellow Vodouisant, Roger, the *oungan*, however, said

Vodou religion doesn't provide an answer as to why the earthquake happened – even though it does have a belief in God (*Bondye*) as creator.

Even accepting the earthquake as a natural phenomenon, many participants also saw the creation of God as despoiled by the sin of humankind. Alcee believed God was creator, master and architect of the universe and that God endowed humans with intelligence to care for this world. However, due to human disobedience things do go wrong in the world. He believed, even though there is a scientific explanation for the earthquake, God allowed it to happen. Antoine and Bernard believed that disasters happen because of humankind's sin. However, they did not think each natural hazard was caused by any specific human sin. They believed these hazards are natural, but it is the human misery that results because of sin. Though she witnessed some terrible scenes when the earthquake struck Petit Goâve, Cecile said that because of the earthquake she now understood God better as a creator; on the other hand it also led her to realise the fragility of humankind "because we could die in the blink of an eye." Alex however believed that even the earthquake was created by God because no one else could create such a phenomenon. Angelique believed that God created the world with natural fault lines, from which come earthquakes.

The predominant impression we gained was that participants had not given a great deal of thought to any other model of creation than a literal reading of the Genesis account. Nevertheless, the beliefs they had formed made a substantial contribution to their understanding of the earthquake and its effects on their country. These creation beliefs held a common and natural place in their narrative of response and recovery, and, because they provided some explanatory framework to a most terrible and terrifying event, they provided essential comfort and strength. However, strong culturally embedded religious beliefs did not signal lack of interest in, or lack of desire for, scientific education in natural hazards and disaster mitigation. In both academic and local community contexts, we were invited to present sessions in natural hazard awareness, at which rapt attention and interest was shown.

Providence

Of all the evidence from the theological influences survivors shared with us, the greater part referred to a doctrine of divine providence in some form or other. However, it is worth noting that there is a cultural nuance for Haitians against which the doctrine of providence needs to be understood. The nuance was explained to us by Theodore, Director of Social Services and Mental Health at the *Hôpital Universitaire de Mirebalais*. When we asked him to what extent spiritual and religious beliefs were considered significant in the treatment of psychological trauma arising from the earthquake, he told us, "If you can talk to any Haitian, you will find in most of them that whatever happened, it was permitted by God." Such a comment can be interpreted, by researchers in natural disasters, to indicate "typical" tendencies toward what is termed religious fatalism.[7] However, Theodore had a much more positive and constructive perspective from working as a mental health professional. He explained,

> And, if it [i.e. the outlook described above] is so, he [i.e. the typical Haitian] *had the answer as well* [as the problem]. For me, it [the outlook] was a rock where I tried to help them to build on to get back. Not from the image of a punishing God, but something happened and he [God] will provide. I remember a women [sic], quite educated, who said in a group therapy, "The one, the one who made the shoulders, it is the same who distribute the crosses." That means God . . . doesn't give more than we can carry. And this is to me . . . something that people can use to promote resilience. And I think it has helped a lot.
>
> (Emphasis mine)

Many of our participants' narratives show they were actually using such a perspective to alleviate their shocking experience of the earthquake. We provide an earthquake-survivor narrative basis for this claim under a number of sub-themes of their overall belief in divine providence in the face of their encounter with catastrophic disaster.

God doing his work

A recurring phrase in the interviews was "God doing his work" (Kreyòl: *Bondye ap fe travay. Li*), and this was repeated across the demographic spectrum. Eighty percent of respondents said that the statement "The earthquake was God's will" was "very true."[8] It struck us that this phrase was a common descriptor for divine providence in the Haitian theological and cultural lexicon. It did seem to provide comfort and reassurance amid the chaos and suffering brought on by the earthquake: in spite of all that was going on, God, in sovereign providence, was going about his work. This was never said to lay blame at God's door or out of terror, but it was meant to suggest that God is not beleaguered or diverted from his purposes; he *uses* even an earthquake to do his work. The belief seemed to create a sense of normality, or of stability that brought reassurance to people in the midst of an event in which nothing else seemed to be normal or stable: not least, the ground they stood on and the buildings they lived in. It also provided a sobering perspective for Christian communities, confronting them with the fragility of life. This belief is very different from fatalism, which classically emphasises the impenetrable and impersonal force of the inevitable happening, with which there is no human free will or moral purpose.[9]

All is in God's hands

Life in the slum district of Cité Soleil can often be a matter of routine survival. Like many residents, some of our participants had migrated there from outlying provinces, in pursuit of employment and vain promises of a better life.[10] Having lost both her parents while living in the provincial town of Jeremie, Antoinette went to live in Cité Soleil in 1994. She lost her partner during the earthquake under the ruined central market place in Port-au-Prince. She and her children spent two years in an IDP camp near the airport where she caught cholera and was vulnerable to gender-based violence (GBV). She later re-located back to Cité Soleil. Although she struggled to pay her rent, to buy food or to send her children to school, she expressed her trust in God's sovereignty, believing him to be the master, the person in charge and capable of doing everything. She would have liked the Government to have found means for her to meet her needs, but given this was unlikely to happen, she stated:

> It's still God who makes it possible for me to stay alive and gives me courage to keep on living. If I'm still living today it's he who makes it possible for me to live. Even when everything becomes very dark for me, I know I've got God who will do everything for me. If I didn't, if I didn't believe in him, then, because of the twelfth of January, I wouldn't be here. . . . All the misery, all the poverty I'm going through at the moment, I leave it in God's hands. It's God who gives . . . I'm in God's hands and at the disposal of God's will. I will always remain firm in my faith in regards to what happens, with regards to what I meet along the road.

The perspective on the political governance of their country by Cité Soleil participants was very dismal. They saw the failure of the Government to have contributed to their poverty and wretchedness. People felt the Government only helped their own people and the armed gangs. Apoline explained,

> Those that have people in the government that found aid, like a friend. But me, I don't have anybody in the government. Here's how it is, you may have your own person (contact) in the government, and you have two or three friends somewhere. You are given from your friend in the government, so after receiving you share it with your other friends.

Alexandre saw a huge disconnect between the people and the Government, with no effective collaboration. People saw their Government working against them rather than for them. Virtually everyone outside the capital complained at disinterest from central Government. Even Maria, mayor of Léogâne, complained. So, in a sense, Aolphe spoke for most people when he said he was certain that there was no real State compassion in Haiti: "The State doesn't look after us, *Bondye* [God] does." Most participants placed more trust in God than in their Government.

Providence and human responsibility

Divine sovereignty and providence, generally speaking, were not seen in conflict with belief in human responsibility. Nevertheless, we found in some cases amongst the rural poor and less educated that a belief in divine providence gave a level of emotional comfort that was prohibiting initiative for development and change that could save lives and property, and enhance quality of life, in the future.

This emotional comfort proffered by the doctrine of providence may be naïve, but in our view any naïveté is simply due to lack of education and information. The naïveté was explained to us in a conversation with Simeon, a Catholic businessperson with a pragmatic and entrepreneurial attitude to life.

Simeon shared with us his deep concern for the people of the country he loved and especially over those social sectors that seem naïvely accepting of their extremely basic living conditions. He spoke of people who live off one cooked meal a day made of very basic natural leafy foodstuffs collected locally. Many of these people would not recognise they were in poverty, he told us, because they do not know anything different. From our experience of listening to many people in such circumstances, we heard them relate their comfort, gained from reliance upon God amid primitive living conditions they took to be the provision of God. The point Simeon was making, however, was that the issue was more sociological than theological. He reasoned that people are content in poverty because they do not know anything else. They accept a traditional social system that maintains them in a poverty they do not even recognise, and they are comforted by their belief in divine providence to remain in that system. Simeon described the dilemma: "Knowing come[s] from knowledge; so you have to know to understand where you are. So if you don't know, you are fine!" He said,

> Those people will tell you, "We call it rich! We are free of bills. Free of pollution. Free of all of the great necessities that you see!" To which we might respond, "Man, you are . . . oh, look at you, you are poor." To which they would then reply, "OK then. I might be poor. But I don't

> owe the bank four hundred thousand dollars. And I sleep lovely. And I eat good. And I'm healthy!"

In one sense, Simeon could see their wisdom, but on the other hand he was clearly frustrated that so many of the Haitian people in rural communities were content in that "poverty mindset." He was frustrated that those people could not see that their lives could be so different if only they had knowledge. However, the lack of knowledge for many Haitians is not their own fault.

Our conversation with Simeon raised another issue related to divine providence and human responsibility. This was over what constitutes so-called development and how that contributes to character and flourishing. Simeon's point was a significant one. Our research in Haiti has shown us something of the real dilemmas facing the rural poor over development strategies and projects. For instance, the mountain road projects at Trol Chouchow received the support of people of all ages living along the routes.[11] They saw, and applauded, the advantages of far easier access to the urban markets for trading and to the city hospitals for healthcare. When we asked them if they worried about the access the roads would also give to people in the cities to come into their area and buy up land and create new communities, no one voiced any objections to us. Also, as the current generations of parents dearly wish their children to find education and to develop in knowledge, and as their children spend more time down in the cities receiving their education, then increasing numbers of young people living in the rural areas were discovering the benefits of education and travel for developing their lives materially and culturally.

There is a major problem, however, over how the new roads could remain sustainable once the short-term INGO programmes had finished. How feasible was it to expect the local Haitian population to be able to maintain the roads for future years, given the annual rainy seasons and hurricanes?

How do you continue to trust in a divine providence that has provided a "good life" for so long, while at the same time entering into a responsibility for developing a way of life that opens up for you far greater prospects for improved healthcare, diet, employment and the improvement of rural agriculture and, not least, the financial gain dependent upon these? It could be that an equal level of dependence upon the providence of God in *conjunction with* the human responsibility for engaging with morally justifiable and sustainable developmental programmes will provide an answer to this dilemma.

Providence – harsh perspectives?

In some instances, we feared the providence doctrine was being utilised in a somewhat individualistic, even selfish manner, especially where an individual's escape from death or injury was interpreted solely in the interests of that person's survival, not with the suffering or death of others in mind. For example, Armand witnessed a large building collapse on other people but it did not fall on him, for which he thanked God. In the centre of Port-au-Prince, Emilie had recently finished her school day and was sitting with a friend in the Champs de Mars when the earthquake struck. The earthquake destroyed her school, killing around eighty-eight students. She then returned to her home and saw it had collapsed, killing her mother and sister. She reflected to us her gratitude for God's providential grace in that had she remained inside her school, or had she returned home too soon, she too would have died. Others gave the distinct impression that they read the mercy of God to them through the fact that they were spared while others died. Certainly, four years after the earthquake, relief over survival among some participants seemed more real than sorrow for the loss of even close relatives.

Another reaction that seemed harsh to us initially was the way participants developed tunnel vision as they made their way back to their homes after the earthquake, even though they had passed many injured people crying out from under collapsed buildings. It occurred to us there was actually some essential intuitive mental triaging taking place within their minds that enabled them to prioritise placing remaining family members first. Desiree told us that she witnessed the school she worked in collapse, killing the majority of the students and staff inside. Whilst she did watch the rescue of some survivors in the immediate aftermath, she soon hastened back home to check up on family members. She found her house partially collapsed but was able to dig a hole large enough to release an aunt who was trapped beneath the rubble. Just after she had done so the whole building collapsed and this would have surely killed the aunt had Desiree decided not to go home when she did.

It would be remiss to conclude from such apparently self-oriented descriptions that people were callous in their reading of divine providence. It is an instinctive survival reaction for those caught up in a catastrophe being focussed on their own survival when many others around them are dying; it is equally normal to want to focus on your own family's welfare.[12] Adolphie, however, put this response down to people being in a malaise of shock caused by them not knowing what an earthquake was.

Further puzzling perspectives arose from other interviews. Pastor Augustin told us that many people had converted to Christianity following the earthquake but had then relapsed. We asked him why he thought this had happened. He suggested:

> Well, maybe it was because of their fear? Because it says that in the Bible before the foundation of the world, God had already decided who

> was going to be chosen, who was going to be elected; that, therefore, there were those people who . . . were predestined to live and those who were predestined to die.

In other words, predestinarian interpretations could have provoked such a fear in some people that they felt the only safety and assurance they could find in the crisis would be to convert to Christianity, to ensure being among the "elect." In view of the awful things Augustin had witnessed from the earthquake, we asked him why he thought God allowed those things to happen to people. He answered that he had never asked that question, but he dispassionately appealed to the biblical references where Jesus announced that such terrible events would happen and that these were signs that it would not be much longer before Jesus would return. This was the closest we heard any participants come to the view that there was an air of inevitability about what had occurred; since Jesus/God had said it would happen it was not something one should become too emotional about.

Many felt that biblical disaster warnings were clear indications of God's providence giving meaning to the events of the earthquake. We must admit that no matter how much some of the participants gave us the impression they did find such interpretations helpful, there were some that left us feeling quite disturbed as to what kind of compassionate contribution their interpretations of providence could offer to other survivors, or even to themselves.

In view of what they must have witnessed from the earthquake, the responses of some participants left us wondering whether they had become completely inured to suffering. A woman in Durisy said she viewed the

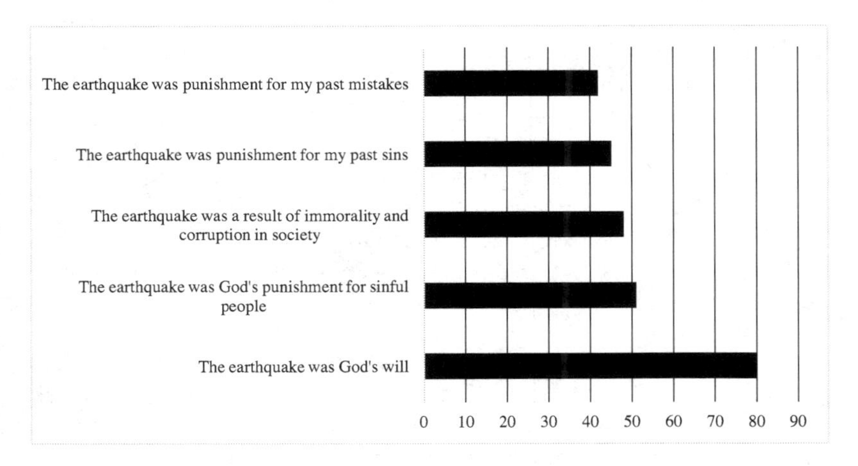

Figure 3.2 Percentages of Haitians indicating that the statement was "very true."

earthquake being the will of God in judgement for humans' sins. Since the comment was made amid some laughter we were not sure how much she really believed this or how much she was just repeating a commonly stated view. Nevertheless, half of our participants agreed that the earthquake was God's punishment for sinful people, or of immorality and corruption in society. Only a small percentage fewer (45%) believed that the earthquake was punishment for their own sins.

It is difficult for us to judge how much this dark perspective is a product of a Haitian ethnographic theology born out of their social history of suffering and struggle, or how much it is the product of the belief systems of a vindictive Vodou culture, or the theology of the mainly U.S. Protestant Evangelical missionary work that entered Haiti during the nineteenth and twentieth centuries. Some indication that the latter is true was provided immediately after the earthquake, when the American Protestant Fundamentalist pastor and broadcaster Pat Robertson, during one of his television broadcasts, brazenly asserted that the earthquake was a direct act of divine judgement on Haiti because of the curse that lay on the nation since the alleged Vodou pact made between the revolutionary slaves and the devil at *Bwa Kayiman*, in the north of Haiti, in 1791. An indication of the abiding strength of this view among the Haitian Protestant population was the organisation of at least one service of national repentance after the earthquake for Haitians to gather and seek God's forgiveness over the alleged *Bwa Kayiman* pact.[13]

Hope

A general perspective on hope, and heard often in general conversation, comes with the Kreyòl proverb: *Koté gen lavi, gen espwa* (Where there is life there is hope). This belief undergirds the historical Haitian determination to overcome, if necessary down to the last breath, never losing hope. From a Christian perspective, "Where there is life there is hope" added significantly to the Haitian therapeutic narrative. What contributed to our participants' voices of hope was their strong belief in God's providence. Hope believes that we are capable of becoming much more than we are now; the life of faith is rich in possibility no matter how apparently hopeless it seems at the time.[14]

We found there to be considerable hope among the Haitian people we met, but hope was rooted in different convictions. For some their hope lay in life after death. For others, their hope lay in the potential for Haiti under different circumstances of governance and infrastructure, as well as in what lay in the life beyond death. For others, hope needed to incorporate education or employment.

Hope in this life

Frederick is a married, educated employee, who had been robbed three times on the streets of Cité Soleil, as he travelled back and forth from his employment. He said that the earthquake in one sense made life even more hopeless because it had led to increasing unemployment, as businesses destroyed in the earthquake had not been able to recover. Yet, on the other hand, in his view, people had turned more to place hope in God; even criminals had begun to do so.

The case of Caroline adds credibility to the point made by Frederik. She grew up in Cité Soleil, dropped out of education and became involved in the notorious gang culture. She had witnessed a lot of violence, even rival gang massacres and one day a particularly horrific slaughtering of a Christian pastor. However, soon after the earthquake, through the influence of a Haitian pastor, she had found work with a Christian INGO and had converted to the Christian faith. Both these experiences had given her the only hope she had for her future – employment and hope after death. She was sure that much of the violent culture was driven by the daily desperation created by the poverty endemic in Cité Soleil. When asked about hope she replied,

> I don't believe there's any more hope for Haiti other than [the INGO she worked for] helping out, and somebody accepting Christ and waiting for death. . . . There's hope 'cause I'm still alive, and as long as I'm still alive I will still have hope. And God has done this thing for me [provided work] that now I am under his care, and so I have even more hope.

Hope in the humanitarian institutions

Haiti has been dubbed "The Republic of the NGOs" for good reason, since around 10,000 non-governmental organisations were in Haiti long before 2010.[15] After the earthquake that figure doubled. Though huge controversy surrounds the achievements of such a vast number of organisations, each with a focus on providing help, the "invasion" did provide some hope for devastated people.[16] We witnessed hope first hand in the INGO medical centres where we conducted some of our interviews. The poverty in Haiti raises the healthcare needs exponentially, earthquake aside. Health and hope are strongly linked. As we conducted interviews in faith-based medical centres, these facilities were havens offering some hope in a country where the earthquake had obliterated healthcare facilities and expertise. Staffed by both nationals and foreigners, they started each day's work with a meeting for worship, prayer and Bible-based ministry for both staff and patients, led by Haitian chaplains. Many Haitian staff there generated immense Christian hope themselves: their Christian faith evoked compassion in addition to the material benefit that paid employment provided for them personally. These centres became at least temporary beacons of hope.

A notable case was that of Jonaissant and of his wife Dominique, mentioned earlier. Roger first met this couple in a Médecins Sans Frontière hospital in Léogâne in early 2011, in his role as an INGO chaplain. Jonaissant was working for the same INGO. Six years after their survival of the Gonaives flood, and just one year after surviving the earthquake and losing their twin boys, Jonaissant had suffered a work-place accident, which he feared would lose him his employment, thus adding to the family's suffering. Roger was visiting him to assure him, on behalf of the INGO, that his future employment with them was safe. In a research interview in 2013, when we asked whether he and his family still lived in hope, he replied:

> In God; only in Christ. If it was for . . . if my life depended solely on material gain I would not have been alive because I do not have any of these things; but I have hope and I know that God is the one taking care of me.

Figure 3.3 Close encounters: life for 50,000 Haitians in an IDP camp in the Delmas area of Port-au-Prince.
(January 21st, 2010)

Source: Photo by Fred W. Baker III, U.S. Department of Defense

A year later, when we found him still living in a temporary shelter, having been laid off by the INGO, Jonaissant repeated,

> Hope that God can do everything for me. Because He is the only one I can count on, He is the only one that I have. I've people [to] take care of me, but it was Him who spoke in that person's heart to do something for me. I count only on Him.

Hope in religious institutions

Amid the destruction in Léogâne, hope could also be found during this life, especially in community with other Christians. Françoise, a pastor, testified:

> You have to have hope because as a Christian you cannot live without hope, and even in the church itself we support each other. There are those that do not have means and so we would buy things for them so as to support them, whether it be food or clothing.

The community provided by his church, and their practical care for one another, brought hope to survivors.

When we located Pastor Otis, at a Christian community near Arcahaie, north of Port-au-Prince, he told us how their mother church, in the Delmas district of the capital city, had ended up in that rural community in the immediate aftermath of the earthquake. Initially, the congregation had remained together at their Delmas site. After some days, as life in the city became increasingly intolerable, the congregation started to evacuate. The evacuation caused a massive logistical and administrative adventure that saw the local churches and communities around Arcahaie rallying together, pooling their meagre resources to help the 2,000 evacuees as they arrived at Otis' church. Even the local Vodou *oungan* came to provide assistance. Summing up the operation, Otis told us,

> And so, when the people came from Port-au-Prince and they came here, so this place was sort of like a place of restoration for them. It is here that those people came and they were restored mentally and physically before going back to PAP [Port-au-Prince]. And that is why a lot, among the refugees that was here, a lot of them always come and visit because they owed this zone this reverence, so they are grateful because they know the good deeds that this community has done for them.

Sadly, the project received no governmental assistance. The whole project was catered for by the Christian churches and by the local community.

Hope for women

Given Haiti's historic record of sexual and gender-based violence (SGBV), echoes of hopelessness would not be surprising if they came from women following the earthquake when reported episodes of SGBV spiked enormously.[17] Even allowing for the media exaggeration and for Western, HICs' cultural sensitivities towards SGBV (also known as VAWG – violence against women and girls), the issue is a serious one in Haiti's history, where rape has frequently been used by dictators, military juntas and by foreign occupiers to subjugate women and to humiliate men.[18]

In 2014, we interviewed Rian, an INGO manager leading a post-earthquake sexual and gender-based violence programme entitled "Restoring Hope and Dignity." She told us that her team had encountered a lot of uncertainty among women since the earthquake, with many not knowing each day whether they would live or die from abuse. The resilience of many women had already been reduced by a culture of SGBV within Haitian society long before the earthquake. During the earthquake, many young girls were left on the streets and exposed to abuse and rape. Rian found that testimonies from the IDP camps said that rapes there usually took place between the hours of 10 p.m. and 3 a.m., including raping children as young as 3–5 years old. Violence aside, the earthquake seemed to destroy hope in these women. As Rian said,

> Sometimes I feel like the earthquake sometimes – it's just how I feel – made them hopeless, made them lose hope in, because they seem to feel that no one cares about them, no nation, no system, no government out there, that cares about them as a nation . . . I'm just telling you what I think, and also from speaking with individuals; I'm talking to people and that sometimes brings them to a point of saying, "Who cares?!" and so [for the] earthquake to come in, it feels as if even God doesn't [care]. And so, I felt like it's, when a woman, tells you, "Oh. I was raped, yeah; not once, not twice, yeah, people just come into the house! And I, my son too, oh my grandchild also!" It sounds more like a resigned place where nobody cares. I mean you should see . . . I mean you should see the places . . . you've seen the places they live in. You've seen the little houses they stay in. You've seen . . . even families, all the Christians that we deal with, it's a husband and a wife and five children in one little room! Children exposed to so much. The parents are exposed to so much.

The SGBV programme of this INGO had succeeded in bringing hope to many abused women, so much so that Haitian beneficiaries included church leaders and some Government officials. The programme focussed on training and empowering the churches to engage men as well as women, to teach them biblical principles of gender respect and to train the women in setting up their own small businesses. To enable women to do this the team had also to teach literacy classes to help in keeping accurate small business accounts. Churches were also encouraged to give out small loans. These were small tools offering hope for abused women. In Haiti, women are the *potiman* (centre-post) of society, culturally mandated to remain monogamous and to take responsibility for income generation and productivity for the home.[19]

We encountered a personal story from a participant that is relevant at this point. It concerns Cecile. High up in some mountains there is a community we visited on numerous occasions. When conducting interviews there we met Cecile, whose earthquake experience we related previously.[20]

It was a bitter irony for Cecile that soon after the earthquake she was raped by a male "friend" while visiting Port-au-Prince, resulting in her becoming pregnant. The man tried to compel her to abort the baby, but she refused; based on her Christian faith, she said that she would rather die with the child than have an abortion. After this response, the man would have nothing more to do with her. Therefore, she went back home to her family and her church, risking the disgrace and shame her condition meant in the Haitian rural culture. Three years later, she sat in our interview along with the child she had refused to abort, the child her faith had given her strength to bear and raise amid the poverty of her mountain family home. Over the research period, it was a privilege for us to talk with this woman and to meet her family. We could see how the social stigma she had felt so deeply initially became progressively overcome, and she felt able to speak with her pastor about her incident and became accepted back into the church life from which she had dreaded being ostracised following the rape.

Hope in the life to come

For some, their hope only lay in life after death. For these people, life here and now held little or no hope for them, so great was their sense of loss and/or of suffering, but not just from the earthquake. The earthquake exacerbated their routine struggles. Their hope lay in the promises of the Bible assuring them of a future life once this miserable one had ended.[21]

A tragic social phenomena in Haiti meant the case of Adelaide was not particularly rare, but if it were not for her faith her case would set a tragic existential context for people like her living in Cité Soleil. Born in an area near the border with the Dominican Republic, Adelaide's mother died at an early age, leaving her in the care of a father who suffered poor health. At the age of 7 or 8, she became a *restavek*. That is, her father gave her into the "care" of another adult couple to become a slave in their home.[22] Her life revolved around washing, cleaning and cooking, also she never received any education to enable her to read or write. This couple would take Adelaide with them to the local evangelical church. However, by the age of 11 the husband of the couple frequently physically abused her. At 12 years of age, she became a servant in another house, in the Delmas district of Port-au-Prince. Sometime later, when living in Cité Soleil, she entered into a *placaj* relationship and had several children.[23] Life in Cité Soleil became increasingly hard before the earthquake. During the earthquake, she witnessed terrible things, including a young girl being crushed to death as the house she was evacuating fell on her. Adelaide's own house was damaged, but she received no help from aid agencies or the church. Then, when the cholera epidemic struck nine months later, her partner died of the disease. This loss left her feeling it would have been better if she had died rather than her partner, since she had no means of supporting her children. As one participant had told us, in Haiti not even when you die do your problems always leave you, or at least leave your dependants! When we asked Adelaide what hope she had in life following such a catalogue of misery, she replied that her only hope was in God and in his provision and care, though she struggled to identify just what that provision could be. Her ultimate hope was that a life provided by God after her death would release her from her lifelong struggles.

Indeed, for so many of the participants in Cité Soleil after the earthquake, if not before, their hope lay in the return of Christ and the future life. They struggled to see any other hope than that, since their situations were so hard in terms of finding work or even finding food. Nevertheless, the hope they expressed in Christ and his return was very positive and strong. It was this hope that they lived and longed for.

Conclusion

The ability to read the shocking events of the earthquake in the way described stems from the Haitian natural character, as Theodore intimated. There is a characteristic urge to seek a final explanation for such events, which is ultimately found in *Bondye*. These interpretations could appear to be those of a religious fatalism that prevents Haitians from realising the importance of human responsibility for disaster mitigation practices and measures. Given the fact that, in their circumstances, their options for survival were extremely limited, making acceptance of their lot being under divine control was a great comfort. After all, no one could have prevented the earthquake itself from happening and, given the condition of the country, people were bound to be seriously injured or to die.

For many, it was the absence of knowledge and information that proved problematic. There is a real thirst and desire for knowledge among the Haitian people. In the absence of such, however, people find the doctrine of God's providence hugely beneficial, but not as a substitute for knowledge.

Those participants who did have higher levels of education did not exhibit a lower regard for divine providence. Their greater knowledge and acquisition of information simply nuanced their understanding of the doctrine with a sense of human responsibility that alerted them to a greater awareness of mitigation practices and measures, and to some anger that their Government had kept this essential information from them. Had these participants not had the biblical knowledge they did have, we are sure the impact of the earthquake upon them psychologically and emotionally would have been far more damaging than it was, and the doctrine accounted for the relatively low levels of lasting trauma found among our participants.

Notes

1 Pew-Templeton Global Religious Futures Project. www.globalreligiousfutures.org/countries/haiti#/?affiliations_religion_id=0&affiliations_year=2010®ion_name=All%20Countries&restrictions_year=2016. Accessed: 20/09/2018.
2 The World Fact Book: Haiti. www.cia.gov/library/publications/the-world-factbook/geos/print_ha.html. Accessed: 20/09/2018.
3 Tracey Kidder, *Tracey Kidder, Mountains Beyond Mountains: One Doctor's Quest to Heal the World*, Kindle ed. (London: Profile Books, 2009).
4 Roger Philip Abbott, *Sit on Our Hands or Stand on Our Feet? Exploring a Practical Theology of Major Incident Response for the Evangelical Catholic Christian Church in the UK* (Eugene, OR: Wipf & Stock, 2013), 125–31; Beth Hedva, "Spirituality and Mental Health: My Learnings on Resilience after Mass Trauma," *Haiti*. Online: http://onetv.com/bethhedva/BHinHaiti.html. Accessed: 21/01/2016; Y. Y. Chen and H. G. Koenig, "Traumatic Stress and Religion: Is There a Relationship? A Review of Empirical Findings," *Journal of Religion and Health* 45 (2006): 371–81.
5 Because Vodou is not a text-based belief system, many of the beliefs of *Vodouisants* were passed on orally or learned through Catholic oral catechesis.
6 "For nation will rise up against nation, and kingdom against kingdom, and there will be famines *and earthquakes* in various places" (emphasis author's).
7 John McClure, et al., "When Earthquake Damage Is Seen as Preventable: Attributions, Locus of Control and Attitudes to Risk," *Applied Psychology: An International Review* 4 (1999): 239–46; John McClure, "Fatalism, Causal Reasoning, and Natural Hazards," in *Oxford Research Encyclopedia of Natural Hazard Science* (Oxford: Oxford University Press, 2016). Online publication (April, 2017). doi: 10.1093/acrefore/9780199389407.013.39; Douglas Paton, "Readiness for Natural Hazards," *Oxford Research Encyclopedia of Natural Hazard Science*

(Oxford: Oxford University Press, 2016). Online publication (December 2016). doi: 10.1093/acrefore/9780199389407.013.2; Benjamin Wisner, "Vulnerability as Concept, Model, Metric, and Tool," *Oxford Research Encyclopedia of Natural Hazard Science* (Oxford: Oxford University Press, 2016). Online publication (August 2016). doi: 10.1093/acrefore/9780199389407.013.2; Gabriele Ruiu, "The Origin of Fatalistic Tendencies: An Empirical Investigation," *Economics & Sociology* 6 (2013): 103–25. doi: 10.14254/2071-789X.2013/6-2/10.

8 Data from 212 surveys reported by Erin O'Connell, Roger Abbott, and Robert White, "Emotions and Beliefs after Disaster: A Comparative Analysis in Haiti and Indonesia," *Disasters* (2017): 803–27 (See also Fig. 3.2). doi: 10.1111/disa.12227.

9 See Mark Bernstein, "Fatalism," in *The Oxford Handbook of Free Will*, ed. Robert Kane (Oxford: Oxford University Press, 2005), 65–81. Abbott, Roger Philip. "'I Will Show You My Faith by My Works': Addressing the Nexus between Philosophical Theodicy and Human Suffering and Loss in Contexts of 'Natural' Disaster.' *Religions* 10, 213 (2019).

10 Jean Casimir, "Two Classes and Two Cultures in Contemporary Haiti," in *Contemporary Caribbean: A Sociological Reader*, vol. 2, ed. Susan Craig (Maracas, Trinidad and Tobago: The College Press, 1982). See also Amy Wilentz on the "American Plan" of the Reagan administration in the U.S., in *The Rainy Season: Haiti Since Duvalier* (New York: Simon & Schuster, 1989), 269–71; Peter Hallward, *Damming the Flood: Haiti and the Politics of Containment* (London: Verso, 2010 ed.), 5–6.

11 However, administration of these projects by the INGOs responsible for their oversight did cause a number of conflicts with the local Haitian communities that provided labour.

12 From our survey 71% said they did not feel guilty for not having done more to rescue people; 18% said they did feel guilty. Gina Ulysse offers a withering exposé of Western media interpretations of traumatised Haitians' apparent callousness in the face of earthquake deaths in Ulysse, *Haiti after the Earthquake*, 27.

13 Previous services, or prayer programmes, had been held in 1997, 1998, 2003, 2011 and 2014. For details of the *Bwa Kayiman* pact see Wilentz, *Farewell Fred Voodoo*, 75–7; also Elizabeth McAllister, "From Slave Revolt to a Blood Pact with Satan: The Evangelical Rewriting of Haitian History," in *Studies in Religion* (Thousand Oaks, CA: Sage Publications, 2012). doi: 10.1177/0008429812441310. Our survey of 250 respondents indicated 51% believed the earthquake was a divine punishment; 42% indicated it was not. However, when asked if they thought the earthquake was punishment for *their own* past mistakes, 50% said they did not believe this was the case, and 42% thought that it was.

14 Abbott, *Sit on Our Hands*, 47, 51–5.

15 Wilentz, *Farewell Fred Voodoo*, 175–7; Farmer, *Haiti after the Earthquake*, 4, 99; Hallward, *Damming the Flood*, 177–82.

16 Farmer, *Haiti after the Earthquake*, 99. See also the interview by Sophie Shevardnadze with human rights lawyer Ezili Danto, "NGOs Masturbating on Our Pain: Haitian Human Right Lawyer, August 17, 2018. Online: www.rt.com/shows/sophieco/422930-un-imperialism-earthquake-human/; Dan Beeton and Georgianne Nienaber, "Haiti: From Original Sin to Electoral Intervention," *Centre for Economic and Policy Research*, February 24, 2014. Online: http://cepr.net/blogs/haiti-relief-and-reconstruction-watch/haiti-from-original-sin-to-electoral-intervention.

17 Ulysse, *Haiti after the Earthquake*, 34–6; “Against Their Will: Sexual and Gender Based Violence against Young People in Haiti,” *Report by Medicins Sans Frontieres*, July 2017; “Anne-Christine d’Adesky and the PotoFamn Coalition,” *Beyond Shock: Charting the Landscape of Sexual Violence in Post-Quake Haiti: Progress, Challenges & Emerging Trends*, UCSB Centre for Black Studies Research, 2013; Claire Mcloughlin, “Helpdesk Research Report: Violence against Women and Girls in Haiti,” *Governance and Social Development Resource Centre*, August 8. 2013: 1–19; Erica Caple James, *Domestic Insecurities: Violence, Trauma, and Intervention in Haiti* (London: University of California Press, 2010), 45–7; Madre, “Our Bodies Are Still Trembling: Haitian Women Continue to Fight against Rape,” Report from The City of New York Cuny School of Law and the Institute for Justice and Democracy in Haiti, January, 2011: 1–31.

18 Hallward, *Damming the Flood*, 284; James, *Democratic Insecurities*, 43, 75, 77.

19 James, *Democratic Insecurities*, 61–2.

20 See Chapters 1 and 2.

21 These views were often derived from U.S. Christian eschatological views: see Billy Hallowell, “Are Severe Natural Disasters Evidence of the ‘Biblical End Times’? Researcher Reveals Americans’ Fascinating Stance,” *Faithwire*, May, 2017. Online: www.faithwire.com/2017/05/02/are-severe-natural-disasters-evidence-of-the-biblical-end-times-researcher-reveals-americans-fascinating-stance/. Accessed: 06/06/2017.

22 *Restaveks* are children sold or handed over to other families to become involuntary child domestic labourers. Lindstrom reported around 300,000 *restaveks* in Haiti in 2012, and several thousand were trafficked to the Dominican Republic (Beatrice Lindstrom, Freedom House, *Countries at the Crossroads 2012: Haiti*, September 20, 2012. Online: www.refworld.org/docid/505c1731c.html). In 2014, the Haitian Government enacted new anti-trafficking legislation to address the *restavek* and child trafficking issues.

23 *Placaj* is a common-law relationship in Haiti, often more common than *maryaj* – legally binding marriage.

4 What we were never told

And I'm sitting in my desk, back at Purdue, and I get the automated email, you know, from the USGS that says 7.2 southern Haiti. I look at the location on the map, I look at the depth, the magnitude, and I, and then I know instantly that it's going to, it's, it's, it's a catastrŏphe.

(Prof. Eric Calais, Scientific Advisor to the UN in Haiti, 2010–2012)

Introduction

The risk of an earthquake occurring in Haiti had been widely reported in the scientific literature for many years prior to 2010.[1] In May 2005, a magnitude 4.3 earthquake occurred with its epicentre around 5 miles south of Port-au-Prince. It caused little damage and raised little attention locally. However, unbeknown to most Haitians, it did cause enough concern for a U.S. diplomat to warn his Government that, "The last thing Haiti needs now is an earthquake."[2]

The 2010 earthquake released a frightening amount of energy. The ground shaking lasted over 40 seconds, though to those caught up in it that time felt like an eternity. Ground accelerations in Port-au-Prince and other places were large enough to throw people off their feet. In fact, the fault that moved was not the 200 km long Enriquillo fault, which cuts right through southern Haiti and had already been identified as a major risk, but two smaller faults in the mountainous region of Léogâne. The faults that broke that day extended across a length of 40 kilometres (22 miles), extended down to depths of 20 km (11 miles), with ground movements across the fault of up to 5 metres (16 feet). The fault movement released an amount of energy equivalent to 10 billion tonnes of TNT, or 1 million Hiroshima sized nuclear bombs.

The level of awareness of the science of earthquakes across the demographic spectrum reflected by our participants was appalling, relative to the extant science and the information available to their Government institutions. This played a significant role in the resulting scale of injuries and

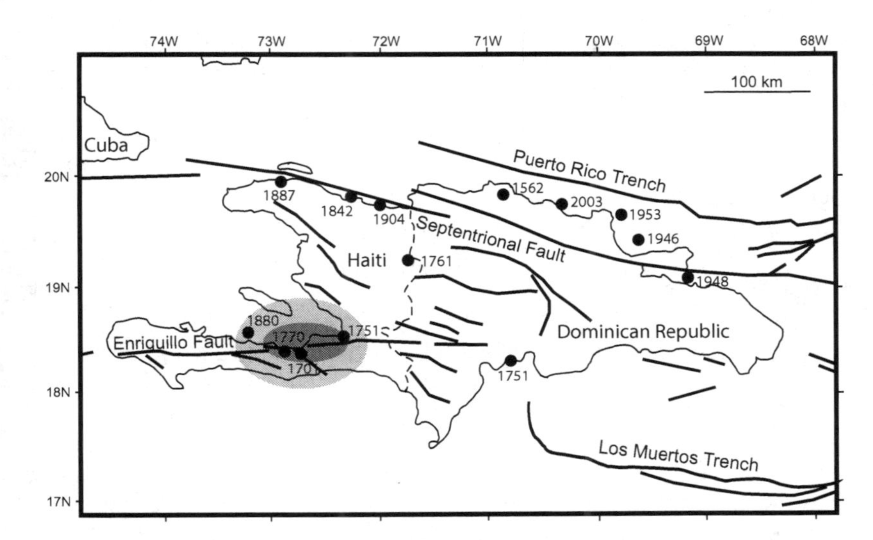

Figure 4.1 Main seismic faults in Haiti and surrounding region. Filled dots show locations of major historic earthquakes, with their dates.[3] Light and heavy shading enclose regions of moderate and severe perceived shaking from the 2010 earthquake.

deaths. On the other hand, the reason for this low level of awareness was, to some degree, exacerbated by the sheer complexity of combatting multiple natural hazards in the region with next-to-nothing fiscal budgets. Such education as there was, coming mainly from the Bible and from the church, was a significant factor in providing a degree of assurance to survivors, contributing to their resilience, as we saw in the previous chapter. Since the biblical knowledge did nothing to prevent serious injury and loss of life, however, mainly being of benefit pastorally in the aftermath, there is still a huge amount of work to be done in Haiti if a future earthquake *disaster* is to be mitigated, let alone avoided.[4] We emphasise *disaster* at this point because earthquakes themselves cannot be mitigated. However, their impact on human individuals and society can, thus preventing a natural event, or hazard, from becoming a disaster.

It was a great surprise to find that so few people had any scientific understanding of earthquakes; even the awareness among the educated middle and elite/governmental classes was alarmingly deficient. Yet they all lived close to major active fault systems. Indeed, a director of a private school said such a state of affairs was, in his view, "criminal." Gaston, based near Cap Haitien, in the north of the country and close to another seismically active fault – the Septentrional fault – thought it outrageous that it took another four years *after* the event for the Government of Haiti (GoH) to inform people that they were living close to major fault lines in both the north and south of their country.

More tellingly, the attitude from within both the UN and the GoH was revealed to us in an interview with Professor Eric Calais, then Professor of Geophysics at Purdue University, Indiana. Calais and his team had been working in Haiti since 2003, measuring earth movements using GPS instruments. From 2010–2012 he was scientific advisor to the UN on reconstruction and mitigation practices.

In 2008, two years before the earthquake, Calais and his team published their conclusions arising from their work on the Enriquillo-Plantain Garden fault. They concluded: "The Enriquillo fault in Haiti is currently capable of a M_w [magnitude] 7.2 earthquake if the entire elastic strain accumulated since the last major earthquake was released in a single event today."[5] This research was conducted in collaboration with the Haitian Bureau of Mines, the Haiti Civil Protection Agency (DCP) and the National Centre for Geospatial Information of Haiti. Its results were communicated to Haitian Government officials and to INGOs in Haiti. After the 2010 earthquake, Calais' observation was that information "on the threat level was available before the event as well as being in international scientific publications, but the overall risk was poorly quantified and not accounted for in building codes, land-use planning, or emergency procedures."[6] Calais also acknowledged the attempts of Claude Prepetit, a Haitian geotechnical engineer at the Haitian Mines and Energy Bureau in Port-au-Prince, to provide some education on earthquake risks. However, these attempts had very limited overall impact, given it requires an army of such educators to inform the whole population.[7]

When we asked Calais how long it took to put their 2008 conclusions into the minds of the Government officials, he gave a fascinating insight into the frustrations that can play out between scientists and politicians. He spoke of meetings in 2007 and 2008 with a number of international agencies, including the U.S. and French embassies, the United Nations Development Programme (UNDP), and even back in 2006, with the Prime Minister of the GoH at the time, Gérard Latortue. When Calais met with Latortue he asked Calais when the next earthquake was going to happen. Calais' reply, as recalled to us, was, "We don't really know, nobody can predict earthquakes", at which point Latortue interrupted Calais and told him that he could predict hurricanes, and he was sure that one would happen that year, between June and October, so he was going to prepare for that, not for an earthquake for which he had no idea when it would happen. Three days after the earthquake the Minister of the Interior, Paul Antoine Bienaimé, called Calais, then back in Purdue, to ask if he could assure the Government of Haiti (gathered in a crisis meeting at the time) that after three days the aftershocks had come to an end. Calais could give him no such assurance. He found a similar response from his collaborative work with the World Bank and the UN.[8] Yet Calais was entirely correct in his response. No one anywhere in the world can predict exactly when an earthquake will occur, though often it is possible to define the areas at risk.[9]

To a degree, given his grasp of the meteorological as well as the geophysical contexts in Haiti, Calais could understand these responses and acknowledged the importance of dealing with hazard response to hurricanes. However, from his perspective as a scientist, the loss of life and economic impact from the earthquake dwarfed everything else. His experience of working in Haiti taught him that whatever was found scientifically, there were many other key players who also had to collaborate (Haitian officials, overly powerful international agencies and donors). This meant that important scientific conclusions often became buried in competing political interests, bureaucratic wrangling and obfuscation. Sadly, his experience also taught him that the intellectual knowledge of earthquakes was just not present within the Government of Haiti, let alone within most of the population.

Education and earthquake risk

Few participants gave evidence of anything above a basic level of formal, institutional earthquake education and understanding. For example, Simon, our research assistant in 2013, was a Schools' Inspector and relatively well educated. He understood that earthquakes can happen anywhere, not just in Haiti; that they are a natural phenomenon; and that if you are living on an active fault line you should realise that at any time there might be an earthquake. This level of understanding was rare. For most, any hazard education, including institutional, was rudimentary and gleaned from various sources, such as the internet, radio and television, post-earthquake seminars and oral information from past generations.

Education

Emile, from Grand Goâve, had heard, somewhere, that there was a geological fault line that ran "all the way to Kenscoff" (a town located in the mountains overlooking Port-au-Prince). He had also heard people saying (correctly) that the earthquake had not been caused directly by that particular fault shifting. Gaston, an educated Haitian student, commented to me that the earth is always moving: "Scientifically, it's moving all the time. We see it or we don't feel it. It's still moving." Some participants acknowledged that they did have some education on earthquakes while at school, but, in most cases, it was exceedingly thin on content and explanation. Caroline said that they always talked about earthquakes and volcanoes at her school, but teachers maintained that earthquakes never happened in Haiti. Clea had heard of tectonic plates but did not understand their meaning.

A few participants received natural hazard education from INGOs after the earthquake, raising awareness and providing basic scientific information. Joseph mentioned Prepetit, the civil engineer who worked for the Office of Minerals in Haiti; he also spoke of another specialist who used to give media interviews explaining that it was impossible to tell when earthquakes will occur but that another large earthquake should be expected in the future. Gabriella said she knew a friend who had been in her church in Carrefour *after* the event when someone had been giving teaching on earthquakes, stating that it is when the earth moves.

It was disappointing to hear the newly appointed Director of a school in the centre of Petit Goâve say that the only educational material he was aware of came from Czech and Italian INGOs *after* the event. Even more disappointing was hearing his response to our inquiry regarding the inclusion of earthquake awareness into the school curriculum and the production of an evacuation plan for the school. He said he did not think this was necessary since it was possible now to trace an earthquake, and, now that they have had the 2010 experience, students know what an earthquake is and how to respond. By contrast, in the mountains, Virgil, a local church pastor and rural school director, explained that he already had some knowledge of earthquakes prior to the 2010 event, so when it happened he knew immediately what it was. He also pointed us to some laminated, clearly worded diagrammatic posters regarding natural hazard awareness, stuck onto the classroom wall, superseding a poster outlining evacuation routes for each class that he had drawn himself, which he had shown Roger on his first visit.

A Director of a private school in Petit Goâve acknowledged that he thought people needed to learn better how to live with seismic risk. However, he felt that this would require a lot of work because "people don't really understand, or want to follow, the science." He said there had been a focus for education about earthquakes immediately after the event, but that this had subsequently lapsed. Others told of learning that the event had been an earthquake from publicity and media announcements *after* it had happened and from seminars arranged by INGOs in their churches. However, irrespective of their knowledge and awareness of earthquakes prior to 2010, many participants across the demographic spectrum said they realised how important it was for the population to become educated in this field.

Previous experience

In addition to academic institutional education, another important source of earthquake education came from generational experience of earthquakes and earth tremors. Estelle, of Jamaican extraction, recalled her mother speaking of the 1692 earthquake on the westernmost end of the same Enriquillo-Plantain Garden fault (which destroyed Port Royal, in Jamaica) and of being made aware as a child that Haiti was a seismically active area. Various participants in the Durisy mountains spoke of having experienced severe tremors years before. There were narratives of feeling earth tremors regularly, especially from people living in the Léogâne and mountain districts. Jean-Paul quipped that in a previous period, when they had experienced earth tremors more routinely, "we used to be happy, when we heard the plates [crockery] hitting each other, banging each other." Householders we visited some miles up the river valley from Petit Goâve said they knew it was an earthquake because they had been used to feeling earth tremors over the years. For all of these references to tremors, what distinguished the 2010 event from anything participants felt previously was their strength and duration.

Given the failure of the educational system and any previous experiences to provide sufficient awareness, there was almost unanimous recognition that now people needed to understand something of the scientific basis for the wisdom of constructing houses in a way required to live safely in a seismically active location. So, when we met with the U.K. Department for International Development (DfID) representative, then based in the Canadian embassy in Port-au-Prince, he was excited by the Haitian Government's commitment to disaster resilience, recovery and mitigation, as evidenced by their impending publication, "Risk of Disaster on behalf of the Haitian Government."[10] There were also encouraging signs that universities in the U.S. were providing scholarships and training for Haitian students in disaster-related skills. By 2016, Syadur Rahaman, in learning from the Haiti earthquake event for his native context of Bangladesh, was also impressed by the progress he found had been made by the Disaster Management Institution and Organization in Haiti, albeit with a way to go yet.[11]

Daft rumours, or what?

Though few people admitted to believing them, it is worth noting that certain rumours were abroad in the public sphere about what caused the earthquake. One rumour was that the earthquake was caused by an American construction project to create a submarine tunnel between Miami and Cap Hatien. Another rumour was that the Americans had been testing a weapon of some sort in the sea, or that they had actually bombed Haiti. To many of us outside Haiti it would be tempting to scorn such tales as born entirely out of ignorance, and reflective of a naïve or even paranoid Haitian mentality. However, when we took into account the history of Haiti's relations with the U.S. and other foreign governments, we could understand these rumours having some credence. For example, Metellus recalled a news item mentioning that the Americans had been training military personnel for coming to Haiti after the earthquake – ostensibly to help. Yet, in the past whenever the U.S. military had arrived in Haiti in numbers, they had ended up attacking the nation, not protecting it.[12] So the large numbers of armed U.S. troops deployed into Haiti soon after the earthquake was bound to cause alarm to the Haitian public. Ideas of their presence being in some way linked to the cause of the event, for Haitians, would not be entirely absurd – even more so, given Haitians' suspicions that U.S. foreign policy is after the rich gold reserves Haitians believe are part of their mineral wealth, as Metellus and Otis told us. For a small, developing nation with a long history of being a pawn in foreign power politics, these suspicions are not ridiculous.[13]

Correlation with faith

Most participants' references to any scientific understanding of earthquakes were also linked to religious understandings. When we asked them what they thought had caused the earthquake, many were more inclined to see it as a divinely controlled event than as a purely natural one. Even those who did understand it as a natural phenomenon held that it was also an event under divine sovereignty. Augustin, from Cité Soleil, saw the earthquake as something natural and not caused by any human interference at all. Others, with some earthquake and even scientific understanding, tended to view it as both naturally caused but divinely purposed: what happened as a natural phenomenon God used for his own divine purposes, and it had been prophesied so in Scripture. This view was put concisely by Henri:

> As a personal definition of the earthquake, I know the earthquake is a natural event. The Bible even warns us about it, and so I know it is something that has repeated back in the past. It repeated in 2010, it probably will in the future. While though Christ himself does control natural disasters and each event and the fact that he allows it to happen means to me that it was in his plan.

For these participants there was no contradiction between natural and divine causation since their theology viewed acts of nature (creation) to be acts of God. Living in Cité Soleil, and with hardly any formal education, Angelique could see that "human beings can't make earthquakes, because they don't have enough power to do that. But the way in which God created the world was that there were these fault-lines, and it's those fault-lines that cause earthquakes." For another example, Aristile said it was a natural phenomenon, but God knew it was going to happen. Many people told us that in their view God allowed the earthquake to happen, even though they also recognised that there was some scientific reason for it, upon which they could not elaborate.

When we asked Alexandre if he thought it would be helpful for people to have a greater scientific awareness of what happens with earthquakes, he replied that it would be if one particularly wanted some scientific analysis; however, for him, the most important thing was to understand how earthquakes relate to biblical prophecy. Even so, he also thought, for the sake of the children especially, that it was important for them to know how to build houses properly. As a local pastor and school Director, Virgil told us that he felt the main part of his education regarding earthquakes had to do with the biblical material, even though he also spoke of the importance of building materials. He fully realised that if houses were built the way they should have been then far fewer deaths would have occurred.

Tsunami – terror

Small, localised tsunamis occurred soon after the earthquake, on the northern coast in the Bay of Grand Goave, and impacting along more than 100 km of the south coast, with a peak near Jacmel. On both coasts, the tsunamis were triggered within a minute of the earthquake occurring, and the maximum wave height exceeded 3 metres (10 feet).[14] The northern coast tsunami was triggered by a landslide, when about 400 metres of the shoreline collapsed some 100 metres seawards causing a tsunami that penetrated 54 metres inland. Two pre-school boys and their grandfather were overwhelmed and killed by the wave on the north coast as they stood watching the sea. These casualties were caused by a complete lack of tsunami awareness, since they made no attempt to escape. There was a similar ignorance among people on the south coast, where fishermen stood filming the tsunami on their phones, whereas Sri Lankan soldiers on duty there immediately self-evacuated, since they remembered the 2004 tsunami in the Indian Ocean and responded accordingly. Fortunately, there were no casualties on the south coast, despite the tsunami penetrating 100 metres inland.

Several people had heightened concerns over the likelihood of a tsunami following a future earthquake. During a staff seminar run by an INGO, in which Roger participated, it was difficult to persuade attendees that the greater danger facing them in Haiti, in the light of the 2010 event, was an earthquake itself rather than a tsunami. Heightened concern among

Figure 4.2 Subsidence at Petit Goâve, which suffered a tsunami.

Source: Photo – Hermann M. Fritz, Georgia Tech

Haitians may be explained by rumoured tsunami warnings some time after the 2010 earthquake. One given to some residents in low-lying parts of Port-au-Prince, so sources informed us, was contrived deliberately by gangs of thieves wanting to evacuate parts of the city near the sea, so that they could then rob the deserted houses. As a result of such malicious rumours, crowds of people did move inland to higher ground. A similar exodus happened in Petit Goâve, according to Raimond, though this was following a tsunami warning a day after the actual tsunami had occurred.

Maria commented that many residents seemed to be worrying about tsunamis in low-lying Léogâne; as mayor, she was more concerned about the statistically much higher risk of flooding from rainfall and the river. She exclaimed, "Everyone is worried about tsunami, but what about [what we are facing] every day; . . . if you pardon me, because tsunami come maybe once every fifty years, sixty years; but here we got raining for [every] five minutes!" However, as Fritz et al. wrote, "The people of Haiti and the Dominican Republic exhibited a complete lack of tsunami awareness despite the 1946 Dominican Republic tsunami at Hispaniola's north-east coast."[15] Sadly, we have to concur with this.

Conclusion

There was a mixture of perspectives among our participants regarding science and earthquakes. Some saw earthquakes entirely as divine events for purposes of judgment/discipline or revelation, while others saw them as only natural phenomenon. The majority conceived causation as a combination; that is, God as creator was the ultimate cause of a natural/creative work. In terms of earthquake understanding, and the effectiveness this had on participants' interpretations of what was happening to them at the time, it must be said that it was information learned from the Bible that came to their minds immediately and which helped them most in the absence of any particular scientific knowledge or understanding. Josette, from Léogâne, explained,

> So when the earthquake itself happened [we] thought that that was that period [end times]. And since the Bible also announces it . . . as to the reason why the earthquake happened itself: the scientific explanation did not know about it, but the biblical part did know about it.

Thus, local churches were significant educators of people we interviewed, although the level of education lacked almost entirely any scientific input. Furthermore, it was the biblical knowledge that gave warning of earthquakes that comforted participants afterwards. This gave survivors assurance that these phenomena were under divine control and were not strange and unknown.

However clear it might seem that education is what is needed most in Haiti to mitigate the risk from future earthquakes, it is by no means that simple. Eric Calais insists that the slow progress in disaster mitigation in Haiti is not necessarily due to lack of good science. In our interview, he referred to two mitigating factors regarding lack of scientific education in earthquakes. The first is "the number that summarises Haiti," which is eighty-four. Eighty-four percent of the Haitian population that reaches university level leaves Haiti and never returns. This leaves a mere 16% of the intellectual capital remaining. The second factor is national and international politics and the way such political agencies orchestrate the actions and agendas of different groups operating in Haiti.[16] These two factors are related, in the sense that with only 16% of its intellectual elite remaining in-country it is not going to be possible for the country's own institutions to take the lead and to distance Haiti from foreign interference.[17]

Even so, there are some encouraging signs that Haitians are now in a better position to be told what they were rarely told before the 2010 catastrophe. Even from our own participants we realise that among the higher educated academics and engineers there is now considerable seismic knowledge and awareness in Haiti, an observation Calais also made during our interview with him. Calais managed to get two Haitian civil engineering students to commence Master's degrees in seismology at Purdue soon after the earthquake. Furthermore it is encouraging to note that since the 2010 earthquake the United States Geological Society (with USAID funding) has been assisting the Bureau des Mines et de l'Energie (BME) in Port-au-Prince to establish a Seismology Technical Unit and the first ever national seismic network in Haiti. This Unit is also engaged in outreach programmes for education in schools and with Haitian officials, and is the authoritative local agency for earthquake-related hazards.[18] An agency that seems to be able to attract Haitian employees to stay in Haiti is The Centre National de l'Information Géo-Spatiale (CNIGS), an organisation within the Ministry of Planning, but which gets most of its funding from sponsors outside Haiti and can therefore pay staff well.[19]

Notes

1 P. Mann, K. Burke, and T. Matumoto, "Neotectonics of Hispaniola: Plate Motion, Sedimentation, and Seismicity at a Retraining Bend," *Earth and Planetary Science Letters* 70 (1984): 311–24; P. Mann, et al., "Actively Evolving Microplate Formation by Oblique Collision and Sideways Motion Along Strike-Slip Faults: An Example from the Northeastern Caribbean Plate Margin," *Tectonophysics* 246 (1995): 1–69; P. Mann, et al., "Oblique Collision in the Northeastern Caribbean from GPS Measurements and Geological Observations," *Tectonics* 21 (2002): 1057–83.

2 The diplomatic memo is at: https://cablegatesearch.wikileaks.org/search.php?q=earthquake&qo=27648&qto=20050531. Accessed: 07/04/2015 cited by Jonathan Katz, *The Big Truck That Went By: How the World Came to Save Haiti and Left Behind a Disaster* (London: Palgrave Macmillan, 2013).

3 Figure redrafted from Bakun, Flores, and Brink, "Significant Earthquakes on the Enriquillo Fault System, Hispaniola, 1500–2010," *Bulletin of the Seismological Society of America* 18–30 with additional historical earthquakes from Roger Bilham, "Lessons from the Haiti Earthquake," *Nature* 463 (2010): 878–9.

4 Haiti's President declared in September 2018 that "the population is becoming better prepared since the earthquake of January 12, 2010" (see *Le Neuvelliste*, September 25, 2018. Online: https://lenouvelliste.com/article/193105/desastres-naturels-jovenel-moise-estime-quhaiti-est-mieux-prepare-quavant-2010. Accessed: 28/09/2018.

5 D. M. Manaker, et al., "Interseismic Plate Coupling and Strain Partitioning in the Northeastern Caribbean," *Geophysical Journal International* 174 (2008): 889–903. doi: 10.1111/j.1365246X.2008.03819.x; Eric Calais, "My Haiti Experience," *Space Geodesy and Active Tectonics Blog*. www.geologie.ens.fr/~ecalais/. Accessed: 20/01/2014.

6 Eric Calais, "Draft White Paper on a National Earthquake Risk Reduction Program in Haiti," http://web.ics.purdue.edu/~ecalais/haiti/documents/seismic_program_DRAFT.pdf. Accessed: June 2013.

Arthur Frankel, et al., "Documentation for Initial Seismic Hazard Maps for Haiti," *U.S. Geological Survey Open-File Report 2010–1067* (2010).

7 Nichola Jones, "Haiti to Improve Quake Preparedness," *Nature*, December 13, 2010. Online: doi: 10.1038/news.2010.670.
8 Pers. comm. Eric Calais, 2015.
9 "Future Earthquake Risk in Haiti: Startling Images of Ground Motion Help Scientists Understand Risk of Aftershocks." www.sciencedaily.com/releases/2010/02/100209152237.htm. The USGS estimates a 5–15% likelihood of a magnitude 7 earthquake on the Enriquillo fault near Port-au-Prince in the next fifty years.
10 The nearest information we could find on this document was reference to the Disaster Management Institution and Organization in Haiti (SNGRD) at http://protectioncivilehaiti.net/sngrd.htm .
11 Syadur Rahaman, "Disaster Management Institution and Organization in Haiti," *Dept. of Disaster Science and Management, University of Dhaka* (2016): 1–27. doi: 10.13140/RG.2.2.11879.85926.
12 The U.S. military occupied Haiti from 1915–1934; Wilentz, *Farewell Fred Voodoo*, 175–7; *The Rainy Season*, 262–4; Hallward, *Damming the Flood*, 257–8; James, *Democratic Insecurities*, 10–11, 52–7.
13 See also Wilentz's reflections on "Haitian's historic wariness about the outside world," in *Farewell Fred Voodoo*, 263–8.
14 Hornbach et al. "High tsunami frequency as a result of combined strike-slip faulting and coastal landslides," *Nature Geoscience*, 3 (2010): 783–788; Fritz et al., "Twin tsunamis triggered by the 12th January earthquake in Haiti." *Pure and Applied Geophysics* 170 (2013): 1463–74.
15 See note 13.
16 Jones, "Haiti to Improve Quake Preparedness."
17 Pers. comm. Eric Calais. See also "The Bishop of Fort-Liberté Deplores the Great Migratory Movement of Young People," *Haiti Libre*, October 12, 2018. Online: www.haitilibre.com/en/news-25807-haiti-social-the-bishop-of-fort-liberte-deplores-the-great-migratory-movement-of-young-people.html. Accessed: 13/10/2018.
18 Pers. comm. Eric Calais.
19 Details of the Centre National de L'information Géo-Spacial may be accessed online at http://www.cnigs.ht/mission

5 Our complicating factors

Tout moun se moun, men tout moun pa menm. (All people are human, but all humans are not the same.)

(Haitian proverb)

Introduction

As authors who readily acknowledge our Christian convictions and our passion for theology and science, it would be satisfying to arrive at this point in our analysis of Haiti earthquake survivors by concluding that religion was the "silver bullet" for their response to and recovery from that catastrophe. However, very little in life is *that* simple, and surviving earthquakes, or any natural hazard if it comes to that, is never *that* simple. Perhaps what makes post-earthquake Haiti a *cause célèbre* is the way her case serves as an exposé of so many complicating factors with which many Low-Income Countries contend when coping with natural hazard events. We found our participants' religious navigation skills continuously hampered by these complicating factors that they had not caused, and over which they had little control.

People often ask us how we cope with listening to so many survivors of major disasters sharing with us the most terrible moments of their lives. In fact, the most difficult part is coping with what we understand of the complicating factors, since these reveal to us just how un-natural disasters really are and how evil they are. In saying this, we are not claiming that disasters are caused by the specific sins of the people who suffer, though Adolphie was sure this was the case for people who had practiced corruption: to him such deaths were "spiritual," meaning payback for past sins. The poor, however, are often the least blameworthy, an important point lost on a number of religious leaders in the wake of the Haiti earthquake and on those who continuously adhere to the stereotypical narratives about Haitians.[1] The Haiti earthquake is the *cause célèbre* because it exposes the role of evil on a generically human scale. Myriam Chancey's riposte to a comment made by the International Monetary Fund's managing director, Dominique Strauss-Kahn, on January 22nd, 2010, when he implied that Haitians needed to "escape their cycle of poverty and deprivation *fuelled by merciless natural disasters*" (emphasis ours), is telling:

> Though laudable in intent, Strauss-Khan's remarks suggest that only natural disasters have had a hand in producing Haiti's cyclical poverty and also that the international community's response is one bound up in a response to what cannot be helped, that is, an act of God. Given the religious rhetoric that enveloped Haiti in the aftermath of the earthquake . . . I have to wonder why the international community's response is steeped in neoreligious ideals of pity or mercy rather than in redressing of political wrongs.[2]

There is a familiar saying among seismologists: "Earthquakes don't kill people, buildings do." The earthquake hazard in Haiti became a disaster because so many structures collapsed with people inside, or beside, them: that and because people did not recognise an earthquake when it was happening, so they did not know what to do. Such a diagnosis may seem very basic. However, if we ask why was such basic knowledge not recognised and implemented by so many Haitians – even across the demographic spectrum, even in the Government of Haiti itself – the answer has to be, "It is complicated!"

In this chapter, we identify a series of human factors that made operating faith and survival complicated. Each of these connects in some way to the political (and religious) wrongs Chancey reckons should be redressed, and to the foundation of poverty which was historically laid and is consistently maintained by Haitian elites and foreign powers.

Education

We have reflected on the role of education specific to perceptions of earthquake risk and disaster mitigation in the previous chapter. We believe this is such an important complicating factor systemically because of its impact upon wealth creation and safe building practices in Haiti. Therefore, the education *system* warrants more reflection, as does the role religion has played in this system.[3]

The systemic problems regarding access to education in Haiti were exposed by the cruel fact that although the vast majority of the population were terrorised by a natural hazard, they had little or no understanding of it or of how to mitigate its impact. This is despite there being at least two major active faults running through the country, which occasionally give due notice of their presence through earth tremors. It is not surprising, therefore, that there are Haitians who feel this withholding of information has been a deliberate political policy by the educated elite to keep the majority populace in ignorance, as Archbishop Isodore claimed. In addition, when white foreign clerics, from countries boasting high education, play on this ignorance in the name of religion – such as the Rev. Pat Robertson did when he announced that the earthquake was God's judgment on Haiti's historic pact with Satan – then the injustice becomes outrageous.[4]

One thing that became clear from our work was the passion for education that exists among Haitians. One senior cleric told us:

> One of the things that gives me hope is that the youth, they love school. And a lot of the parents and relatives of a student, they would even sell their own clothing, and they would rather walk bare foot, so they could have money, so their kids can have an education.

Many people spoke of either their parents working hard to support them through education or themselves, as parents now, doing the same for the next generation. From what we regularly witnessed, there is no mistaking that ordinary Haitians value education and they are prepared to hike or ride for miles to get it, in the cool of the early morning or the noonday heat.

Yet we found systemic problems in making education accessible, despite the Haitian Constitution's, the Government's and the international community's theoretical commitments to providing education for its citizens.[5] Class and religion have also played some dismal roles in denying education to Haitian citizens.

At the time of the earthquake, a mix of State, church and private sector schools provided education in Haiti. The greatest proportion (80–90%) was, and still is, provided by foreign missions and churches in Haiti, largely funded by foreign donations.[6] The Roman Catholic Archbishop of Port-au-Prince informed us that the Catholic Church is the largest non-State provider of education establishments. Sixty percent of these schools are in rural areas. Because it is the State church, since 1913 the Government has committed funding to the church for opening and operating schools in rural and poor areas. However, such funding has been sporadic. There are also numerous Protestant schools, run by Haitian Protestant denominations, often funded by INGOs and foreign, mainly U.S., mission groups. Some of these are unregulated. Seventy percent of Haitian schools lack accreditation by the Government of Haiti.[7]

The 2010 earthquake is the most destructive event any country has experienced in terms of deaths as a proportion of the nation's population.[8] The

Haitian Ministry of Education estimated that 4,992 schools were affected by the earthquake – 23% of all schools in Haiti. Of these, 3,978 – 80% of the affected schools – were either damaged or destroyed and therefore were closed after the quake. Eighty percent of school buildings in Port-au-Prince were destroyed, as were the National Ministry of Education and Professional Formation offices, with the loss of school records. Sixty percent of the schools in the *Soud* and *Ouest* departments were destroyed or damaged. Most significantly, and tragically, an estimated 40,000 children and over 1,000 teachers were killed. Furthermore, an estimated 302,000 children were displaced to other departments, with another 720,000 affected children remaining in their home communities. At the peak of displacement, around 2.3 million people were out of their homes, and many became separated from their usual schools, assuming those schools still functioned.[9] A number of students would not return to their original buildings out of fear of entering a concrete structure. Joseph told us that Haiti lost many educated people in the earthquake, either through their death or by their leaving the country.

Classism has meant many could not get beyond primary education, even if they got that far, and many do not get that far, due to lack of money. Often these students would have to leave school and work to help provide money for the household to have a basic living. This, historically and up to the present day, has dogged the development of education for the majority in Haiti. For over fifty years from the day of gaining independence the Catholic Church churlishly refused to help Haiti, although ironically it now claims to provide the best education system in Haiti. Kings Petion and Christophe, in the nineteenth century, set up systems for both primary and secondary education, but these systems were steered towards benefitting the urban elite class.[10] So when the Catholic Church, under the Concordat of 1860, finally agreed to establish a system, it was already tuned to serve the needs of the urban elites.[11] Such connivance to maintain poverty through the exclusion of the poor from education – the key to opening the door out of poverty – has prevailed right up to this day. In Haiti you get what you are prepared, or able, to pay for.[12] In a country that is strapped for cash and dependent upon foreign finance there is little national or international political will on the part of the ruling classes to distribute wealth evenly.[13] Money (or a gun) is power in Haiti. As the Haitian proverb puts it, "*Tout moun se moun, men tout moun pa menm*." (All people are human, but all humans are not the same.)

In the rural areas, where fertility rates are highest, despite an undoubted desire by parents to have their children educated, it is often not affordable.[14] Parents may start their children in school, but if their crops fail, then they may not be able to continue paying school fees and those children have to stop their education, unless the school Director is prepared to give parents more time to pay. One survivor expressed her plight in this way:

> The kids cannot stay in school the way they should be, because sometimes the directors will send them back home because I cannot pay the full amount; and the reason that is, [is] because if I plant a crop – we have what's call "wind" – sometimes water and hurricane destroy part of it, and I can't get the full amount of what I invested in the crops.

Corruption has also had a significant impact on the Haitian education system. Dubious disputes over land ownership sometimes interrupted the ability to function of institutions built on the land. The Haitian Academy University at Lafiteau, north of the capital, is a case in point. The Academy was the brainchild of two enterprising educators who moved to Haiti from the Brooklyn district of New York, after the downfall of the Duvalier dictatorships. In June 2014, the ownership of the site – already suffering from funding difficulties – was contested, under orders of a Frenchman armed with a dubiously obtained, and false, court order. He had had his eyes on this site for himself for a long time. This is just one example we came across personally of the debilitating impact corruption in the elite society inside and outside Haiti can have on the systems of education that are statutorily mandated to ensure access for all.

Marcos told us that the foreign mission he works with refused to be a part of the typical schools sponsorship business that proliferates in Haiti, which can be lucrative for corrupt educators and even pastors. "Orphan" children are used to attract sponsorships for their education, but then pastors pocket the money for themselves. Marcos claimed that such corruption

> is an epidemic here, and that's why I am not affiliated with any mission, that's why we started our own. Now I'm not going to say everyone is like that, but I'm gonna say the majority, the ones that I been in, I'm gonna tell you, eight out of ten, that's how they do business.

Marcos' story echoed features of Haitian education systems that Timothy Schwartz drew attention to in his ethnographic study in northern Haiti in the 1990s. He concluded that the lack of accountability in financial matters regarding NGOs and their financial "support" for schools in Haiti was seriously corrupt.[15]

In the city slums, such as Cité Soleil, the gang culture compounded unemployment and impeded educational progress. Apoline, for example, could not continue at school after thieves persistently stole her mother's takings while she was working on the streets of Cité Soleil as a *machann*. Pierre had benefitted from the financial support of two Canadian Catholic priests who were paying for his education. When they retired back to Canada, the remaining priest refused to pass their money to Pierre and kept it for himself. According to Caroline, it was the gang wars in Cité Soleil that prevented school attendance. However, in her case she chose to opt out of attending school in preference for the gang culture. This choice is quite common among the youth of Cité Soleil, where lack of finance for supporting their education, leading to the inevitable prospects of unemployment, makes the temptation for a life of "easy money" from crime too attractive.[16] A 2008 report by Save the Children estimated that on average 76% of children between 6 and 15 years old were out of school in the five slum areas of Port-au-Prince. This was because they had opted out, not been forced out.[17] The earthquake only reduced incentive for education in the wake of so many destroyed schools, with many students dead and with zero employment prospects. For too many in Cité Soleil, the key to the door out of poverty is in being a member of the gangs which some members of the elite class and politicians hire to conduct their nefarious business.

Employment

The capacity to gain employment in turn provides the means to rent or construct a property in which to live safely. However, the situation in Haiti regarding both employment and construction is complicated. It is also a political tool, as the necropolitics (politics of subjugation to death) of the Duvalier years demonstrated so well.[18] Beyond simply mentioning it, we are not even going to begin an attempt at unravelling the issue of land tenure in Haiti, perhaps *the* most complicated of the complicating factors, which sits as a historic foundation, influencing where and how safely most Haitians can live, work and relax.

Just working out what constitutes "employment" in Haiti is complicated, since there is so much part-time working, casual sales on the street and child labour.[19] Ten of our 158 participants were unemployed. Others worked in finance or accountancy, as psychologists and social workers, in engineering and construction, in the leisure industry, in agronomy and in secretarial work. Three were qualified schoolteachers, one of whom was also a schools Inspector. A few taught in universities and in seminaries. One was a civil servant for the Ministry of Interior, one the *Archeoveque* (Archbishop) of Port-au-Prince. Others either had relatives or were themselves employed in various trades including carpentry, masonry, welding/metal work, jewellery, mechanics, electronics, cultivation, cookery, cleaning, sewing and public transportation. Several had multiple skill sets, a factor encouraged by the World Bank's advice to Haitians after the earthquake.[20] Some worked in the informal economy or made a living off their own initiative.[21] Others found employment with forestation projects, financed by INGOs, in "cash for work" and construction programmes in the wake of the earthquake. Some, such as Ronel, were self-taught, skilled artisans, using the Iron Market or the Apparent Project in Port-au-Prince to sell their products.[22] In the urban centres, people worked as *machanns*, buying and selling products in the street markets, while in the rural mountains, participants worked as cultivators and as *nàs* weavers as well as *machanns*.[23] Our survivors told us that employment in Haiti is tough to find; even when it is found, it remains precarious, especially when an earthquake *katastwof* (catastrophe) comes along. Yet without stable jobs and income, the country cannot develop its disaster resilience and mitigation capacity adequately, nor can ordinary citizens construct properties able to withstand earthquakes and hurricanes. Stable incomes would help finance education and building compliance to seismic and hurricane resistant standards.

Our participants identified a number of factors that made employment in Haiti unstable and precarious for them and others. Antoine was an example of the policies of centralisation, which began under the Duvalier dictatorships (1951–1986) and were further promoted during the 1990s by U.S. foreign policy and Haitian business elites.[24] Antoine was born in the Central Plateau region where his parents had grown sugar cane and maize and had reared chickens. When he was 8, he and his older brother moved into Cité Soleil to look for work. Life became very hard. He recalls bathing in and drinking the water in the gutters. Even now, he has no consistent work but makes a little income as a taxi driver. Most unemployed people in our survey came from Cité Soleil. Apoline explained that her father had never found formal employment since moving to the capital. The nearest he ever got was working unpaid for the local mayor's office.

Haiti's weather affects employment. During rainy periods, hurricanes or drought, crops may be ruined and life made economically difficult. Heloise lived on the seashore of Trol Chouchow and used to sell fish but was currently out of work. She said her parents grew crops, but when the weather was not favourable, they resorted to making charcoal, which they sold in order to make enough money to live off. Of course, this involves the cutting down of trees, which can lead to denuding the soil's value. This poverty cycle, a common feature in Haiti, exemplifies the desperate self-inflicted wounding of some survival tactics among the rural population.[25]

The earthquake itself was a mixed blessing as far as employment is concerned. Some, ironically, because of the earthquake, were able subsequently to find jobs with INGOs. The impact of the earthquake on people already in employment was varied, with some able to resume work soon after it and others being either delayed for a long period or not able to return to work at all. Gabriel was in the former category. Employed as a credit agent for a bank, he received a message from his office in Port-au-Prince asking him to report for work just six days after the earthquake. Some people sustained serious or disabling injuries in the earthquake that prevented them returning to their previous, or indeed to any, formal employment. This was the case with Martin, who lost both hands when his college collapsed on him. Alexis, who lost his leg to falling glass, found himself getting discouraged and without hope because he was not able to find formal employment after the earthquake. Ronel became disturbed because the tremors terrified him so much he could not settle to work afterwards: he was too scared to sit down to work for fear of feeling the vibrations again, so he found temporary work as a taxi driver instead.

For the poor, the earthquake could be ruinous. Angelique used to sell and buy produce as a *machann*, but the earthquake destroyed all her products and her sole source of income, since her husband had no work. Even as a mason, Metellus could not find work in the reconstruction industry after the earthquake. In his view, it was easier to get work before the earthquake than afterwards. Rafael, an electrician, found he had a lot of work immediately following the earthquake, but subsequently this tailed off drastically, leaving him on the point of looking for alternative employment.

Whereas in some cases the cessation of employment was due to genuine damage to the employer or the facilities, in others it was more a matter of employers taking advantage of the crisis to lay off staff. Armand had not gone into work for a few days after the earthquake because rubble blocked the roads. He was unable to contact his employer to explain because the telephone network was down. In addition, his mother died and, as an only child, he was responsible for arranging the funeral. When he did return to work a week later, he was fired for not having turned up the previous week.

Employment was also deeply affected by the invasion of INGOs into Haiti and their strategies for relief and development, both prior to and in the wake of the earthquake. These issues have been well documented by others.[26] Another complicating factor for employment was provided by the Government's tax system, which, under the self-interested influence of the policies of other nations, often penalised the ordinary Haitian worker.[27]

Historical and contemporary politics

All of the above complicating factors were, in one sense, symptomatic of the complicated Haitian politics. Political shenanigans played key roles in the issue both before and after the earthquake. There are two dimensions to the politics of Haiti – national and international/foreign. Both are historically enmeshed. Both have contributed to the negative image Haiti has been given internationally, which, in turn, has influenced investment in the country.[28]

Though very few of our participants said that violence had been committed against them personally by political regimes, Haitian people have been subject to appalling bouts of violence and abuse historically by both national and foreign aggressors. The impact of this on employment has been subtle but real, forming another factor causing vulnerability in the Haitian workforce. Following the ousting of President Aristide in the early 1990s, his supporters, mostly from the poor class of pro-democracy Haitians, dubbed *Lavalasiennes*, were subjected to a period of brutal assaults from the military junta's sponsored thugs (FRAPH – Front Révolutionnaire pour l'Advancement et le Progrès d'Haiti), aided and abetted by the CIA and designed to maintain U.S. business interests above the democratic wishes of the majority of Haiti's population.[29] These forms of violence specialised in attacking the productive capability (as well as reproductive capability) of both men and women. Rape and amputations were carried out on a large scale, reducing the employability of victims.[30]

There is no doubt that the import of disaster capitalism, ostensibly as aid to the reconstruction and development of post-earthquake Haiti, has had a negative effect on employment.[31] The neoliberal trade policies endorsed by former U.S. President Bill Clinton made Haiti remove tariffs on imported rice from the U.S. This advantaged the Arkansas and Miami rice growers' excess produce and undermined the native Haitian rice market that could have maintained the Haitian population in self-sufficiency in staple rice. The result was severe damage to the agricultural sector in rural Haiti, the major contributor to the Haitian economy at that time.[32] One report suggests that the result of World Trade Organization and CARICOM trade policies resulted in the loss of 831,000 agricultural farming jobs.[33] The trade embargo imposed on Haiti by the U.S. between 1992 and 1994 destroyed the farming industry, with the loss of over 200,000 jobs. Ironically, this loss came from the factories that suffered most under the trade embargo.

These political moves plunged thousands of rural Haitians into unemployment as their farms closed and they migrated to the increasingly overcrowded urban slums to look for work, lured by the illusory promise of jobs in the assembly export industries also endorsed by the Clinton administration. Migration, plus a burgeoning birth rate, amounted to an unsustainable burden upon the capital, Port-au-Prince, to provide employment for a city population that grew from 753,000 in the early 1980s to over 2 million at the time of the 2010 earthquake. Given the inadequate infrastructure and building methods for such a population explosion the devastating loss of life from the earthquake must be understood, at least in part, as a product of neoliberal trade policies forced on the Haitian nation by foreign exploiters and endorsed by the self-serving political and commercial elites of the Haitian State.

To add insult to injury, on March 10th, 2010, Bill Clinton, then serving as the U.N. Special Envoy in Haiti, admitted, when testifying before the U.S. Senate Foreign Relations Committee, that the policies he had introduced previously were wrong and had failed the Haitian people.[34] Just two weeks later, Clinton and Haitian Prime Minister Jean-Max Bellerive sanctioned the Interim Haiti Recovery Commission (IHRC). This was conceived by the U.S. State Department and was made up, in the main, by U.S.-friendly financial institutions, such as the World Bank, the International Monetary Fund, the Inter-American Development Bank and the major donor countries (Brazil, Canada, France and the U.S.), but with only minimal roles for Haitian representation. Many Haitians regarded the IHRC arrangement as a sell-out by the weak Préval Government, ceding Haitian authority to a foreign power. The IHRC allowed the continuation of the policies Clinton had "repented" of, with him as co-chair. This sustained the high unemployment within Haiti, or at very best sustained a large cheap labour force to serve the assembly export business.[35] There is a significant risk that these neoliberal policies, which promote individual profit (which is supposed to "trickle down," replacing the collective, communal responses traditional to Haitians), create increased vulnerability in the face of natural hazards in Haiti.[36]

Nadeve Menard, a member of the *diaspora*, speaks of "disaster capitalism," where each disaster benefits the financial market. "Food for work" programmes often use U.S. rice; the urgent work after a disaster provides job opportunities, but they are often given to foreigners; aid workers are required, but they are supplied by INGOs, not Haitians; and relief projects were often imposed on Haiti, but were not invited and not specified carefully to Haiti's needs. Menard writes of inept leadership that pandered only to foreign interventions and to Haiti's elites: "The idea seems to be to turn Haiti into a vast pool of unskilled labor, which will of course benefit the disaster capitalists."[37] Alternatively, as Beverley Bell put it, "One man's disaster can be another man's profit."[38] So, with the post-January 15th, 2015, Martelly-rule-by-decree and with the World Bank giving notice it will not recognise a formal complaint made to the Bank by Haitian farmers in the north (who felt aggrieved by incursions onto their land by foreign, World Bank-supported mining companies prospecting Haiti's mineral wealth), it does not auger well for political transparency or for rural income generation helping to provide natural hazard-safe constructions.

In the complicated field of politics, the egregious practices of neo-colonialism, hegemony, racism and socio-economic injustice, covertly hidden within Haitian commercial elites and the foreign policies of other nations, ensured that the majority of Haitian citizens would not be living or working safely when the earthquake struck.

Construction

The link between education and construction was pressed home by Adolphie when he said,

> We need also, like, how to give a scholarship to Haitian people; like, to the students study at college and tell them, "Okay, we send you, like, to this country. They will teach you how to build your country another way."

By "another way" he meant without concrete, because it was the concrete that killed so many. For Adolphie, it was neither Vodou or the earthquake that was the real enemy, it was concrete.

Some 73% of buildings (*logements* or housing units) in Haiti are one storey, though this rises to 79% of buildings in rural areas. Eighty-two percent of the housing has tin-sheet roofing, but most multi-storey houses and apartments have ceilings and roofs made of concrete (71%). The walls of 90% of buildings in Haiti are constructed out of either cement/cinder block, earthen materials, clisse (intertwined sticks, twigs and branches), or bricks

Figure 5.1 Comparison between collapsed three-storey Turgeau Hospital and the barely damaged thirteen-storey Digicel mobile phone company building, both constructed in 2009 immediately prior to the earthquake.

Source: Photo: Patrick Paultre, reproduced with permission

and stone. In rural areas walls tend to be of earthen materials, whereas in urban areas they are more commonly made of cement/cinder block. Sixty-five percent of rural buildings have floors of clay, while in urban areas they are mainly of concrete (72%).[39]

Two reports on earthquake damage concluded, respectively, "The massive human losses can be attributed to a lack of attention to earthquake resistant design and construction practices, and the poor quality of much of the construction"[40] and, "The lack of building code and standards for the design of structures, as well as the fact that seismic forces were not considered in the design of most buildings explains the failure of so many engineered structures."[41] A simplistic solution might conclude from these that collapsing structures were the major cause of death during Haiti's earthquake. However, the prohibitive economic cost of conformity to appropriate construction codes is a complicating factor. Paultre et al. concluded, "The key objective for future safety remains education and a strong and sustainable science and engineering foundation in Haiti, so that Haitians themselves can take a lead role in designing solutions *adapted to their economy and society*."[42] Many Haitians are forced by their economic status to engage with the informal construction sector, which, by definition, lacks Government oversight and regulation. The damage and destruction caused by the earthquake, creating a high demand for urgent repairs or new construction, has led to many rebuilding in an unregulated manner and to a standard that places them at high risk in the event of another earthquake of similar strength: at some point such an earthquake is inevitable. Among our participants were two entrepreneurs in construction. Simeon and Yve, both from Léogâne, were angry at the way the Government and the INGOs ignored their contributions of para-seismic model housing. Simeon had rebuilt his own home and his hotel, using reinforced styrofoam infilled with rebar and concrete walls with a light-weight styrofoam roofing. Yve showed us round his majestic model house and admitted that even his model required a doubling of the cost per square metre, from $500 to $1,000. He said that the cost of cement had also increased since the earthquake.

We asked Yve if he thought it was possible for ordinary Haitians, with whatever means they have, to build safely. He said he was sure it was, but the Haitian mentality was such that when people start with a small house they forever want to add to it. Therefore, they build an initial foundation and walls that are insufficient to support an additional floor, even though they fully plan to add one later. He had observed from collapsed buildings that the problem was people constructing ceilings/floors that were far too heavy, then collapsing with the earthquake and pancaking onto each other.

Caught in this web of complicating factors that prevented many survivors having the skills and financial resources for educating or financing their disaster awareness and mitigation, they still have to live somewhere. They will therefore choose, or take, whatever is affordable for them – or least unaffordable if they can raise enough credit – to provide shelter for themselves and their families. This "no brainer" seemed to be lost when it came to another "brain-child" of the Clinton Foundation in Haiti.

In July 2011, newly elected President Martelly and Prime Minister Bellerive, with former U.S. President Bill Clinton inaugurated "Expo," the Housing Exposition for exhibiting sixty exemplar, model homes. The Interim Haiti Reconstruction Commission (IHRC) had wholly approved of the idea. The cost of this exposition drained at least $2 million from the public purse for Haiti's reconstruction, including contributions from the U.K. and from foreign and Haitian architecture firms who contributed $2 million more. President Martelly used his inauguration speech to publicise his scheme for acquiring small mortgages, for the benefit of the middle class of Haiti.

A year later, the consensus in Haiti, certainly from a grassroots perspective, was that the "Expo" was a complete failure. The model houses lay empty and vandalised, and the site overgrown with weeds. Models were too expensive for the majority of displaced Haitians. There were suspicions that they represented attempts to provide foreign architecture and construction companies with lucrative business contracts, though some companies claim

Figure 5.2 Collapse of poor quality housing in Port au Prince.

Source: Logan Abassi, United Nations Development Programme, CC BY 2.0

they just wanted to help in reconstructing Haiti. Figures of between $21,000 and $69,000 per housing unit were cited, placing them well out of the range of a lot of INGOs, let alone the average Haitian.[43] It seems that pressure on Government to produce houses quickly for the huge, and increasingly politically embarrassing, numbers of displaced people, prevailed and the whole Expo project got "kicked into the long grass," to the dismay of many Haitian and foreign experts who had given time and money to it.

If we add the complicating factor of corruption again to the mix, then the cost of housing for anything but a shack just escalated. Ambraseys and Bilham aver that the corruption is endemic within the construction industry in the form of bribes to subvert inspection and licensing processes, as well as complicity in cost-cutting, quality-compromising practices.[44] Death figures from earthquakes globally continue to rise alarmingly. Even so, death reduction from implementation of earthquake-resistant designs can benefit earthquake-prone countries, but only those "that have the wealth and willpower to mandate its use."[45] Haiti does not have such wealth, even if it had the political will. Bilham lists three factors responsible for high death tolls from earthquakes and that also prevent the lessons of earthquake engineering being applied. These factors are corruption in the building industry; the absence of earthquake education; and the prevalence of poverty.[46] Again,

Figure 5.3 Ruins of Cathedral, Port-au-Prince.

Source: UNESCO, CC BY-SA 3.0 IGO

each of these is relevant to the way Haiti suffered from the 2010 earthquake. Ambraseys and Bilham's comment is sobering for Haiti: "The structural integrity of a building is no stronger than the social integrity of the builder, and each nation has a responsibility to its citizens to ensure adequate inspection."[47] The international community also must play its part in this responsibility, especially for a country like Haiti, which has suffered much from the interference and abuse of foreign actors' policies in sustaining Haiti as a poor, dependent nation. All policies contributing to sustain Haiti's levels of poverty should be held accountable also for the terrible cost in life and limb from the structural damage and destruction resulting from the 2010 earthquake. Such policies also prevent access to safe building practices for mitigating damage from future earthquakes.

Conclusion

Though we conclude that religious beliefs were of undoubted assistance to our participants' experience of the earthquake disaster in terms of their psychological and emotional recovery, there are endemic within Haitian society many complicating factors which made the outworking of their faith severely challenging. Whereas survivors could live with the fact that the earthquake, as a natural process, was entirely under God's control and order, accepting and living with the complicating factors was incredibly hard. They saw no evil in the divinely ordained natural hazards, but they saw much evil at work in the complicating factors. These mainly involved systems of structural evil that prevented their access to education, which in turn limited their access to financial means, advantaged the elite's and the politicians' agendas for self-aggrandisement and played fast and loose with safe building practices, or placed them beyond the reach of the ordinary citizen. Since almost all of our participants were not involved in either the elite class or the political process, they were severely limited in terms of how they saw their faith could lead to substantial changes in their life with regards to their vulnerability to another earthquake of similar magnitude to the 2010 event. Praying and trusting in the providence of God was all most could do, while through civil activism just a few sought to advance the principles of their faith in practically addressing the egregious political and commercial systems.[48]

However, we believe that the generational experience of a history of struggle in navigating and coping with life under such complicating factors as we have identified above has created strategies for survival in Haiti that served our participants in equal proportion to their religious beliefs, but which would have been much less beneficial without those beliefs when the earthquake struck.

Notes

1 "Pat Robertson Says Haiti Paying for 'Pact to the Devil'," *CNN*, January 13, 2010. Online: http://edition.cnn.com/2010/US/01/13/haiti.pat.robertson/index.html. Accessed: 10/05/2013. See also, Elizabeth McAllister, "From Slave Revolt to a Blood Pact with Satan." and Ulysses, *Haiti after the Earthquake*.
2 Chancy, "A Marshall Plan for Haiti at Peace," 200.
3 Ketty Luzincourt and Jennifer Gulbrandson, "Education and Conflict in Haiti: Rebuilding the Education Sector after the 2010 Earthquake," *United States Institute of Peace Special Report*, August 2010. Online: www.usip.org/sites/default/files/sr245.pdf. Accessed: 05/10/2018. It is a pity that they do not draw attention to the benefits of education for hazard awareness and disaster mitigation in their report, except in regards to the agenda for youth, sports and civic action (see "Education and Conflict in Haiti," 4).
4 See note 1.
5 Luzincourt and Gulbrandson, "Education and Conflict in Haiti," 1–20; Albin Krebs, "Papa Doc, a Ruthless Dictator, Kept Haitians in Illiteracy and Dire Poverty," *New York Times*, April 23, 1971. La République d'Haïti / Republic of Haiti 1987 Constitution de la République d'Haïti. Basse De Données Politique des Amérique. Online: http://pdba.georgetown.edu/Constitutions/Haiti/haiti1987.html. Accessed: 13/02/2015; "Final Report of the National Survey of Catholic Schools in Haiti," A Report by The Episcopal Commissions for Catholic Education, Catholic Relief Services and The University of Notre Dame (June, 2012): n.p.; "Our Goal: Education for All in Haiti," World Bank: Latin America and Caribbean. Online: web.worldbank.org/WBSITE/EXTERNAL/COUNTRIES/LACEXT/0,,contentMDK:21896642~pagePK:146736~piPK:146830~theSitePK:258554,00.html. Accessed: 18/02/2015.
6 Gabriel Demombynes, et al., "Students and the Market for Schools in Haiti," Policy Research Working Paper 5503, The World Bank Latin and the Caribbean Region Poverty Sector Unit, Education Sector Unit (December 2010): 2–37. This uses data from the Demographic and Health Surveys of 1994–94, 2000 and 2005 and the schools census of 2003.
7 Luzincourt and Gulbrandson, "Education and Conflict in Haiti," 3.
8 E. A. Cavallo, A. Powell, and O. Becerra, "Estimating the Direct Economic Damage of the Earthquake in Haiti: IDB Working Paper Series No. IDP-WP-163," Inter-American Development Bank, Washington, DC, 2010.
9 *Office of the Secretary-General's Special Adviser on Community-Based Medicine & Lessons from Haiti*. Online: www.lessonsfromhaiti.org/lessons-from-haiti/key-statistics/. Accessed 23/11/2014.
10 Dantes Bellegarde, "Alexandre Petion: The Founder of Rural Democracy in Haiti," *Caribbean Quarterly* 3 (1953): 170–1.
11 Michel-Rolph Trouillot, *Haiti: State against Nation, the Origins & Legacy of Duvalerism* (New York: Monthly Review Press, 1990), 51.
12 Paul Farmer in *Haiti After the Earthquake*, p. 43, refers to lottery schools (*lekol bolèt)*, because the quality of education was a gamble.
13 As an example see World Bank Country Study, "Haiti: Policy Proposals for Growth," Report No. 5601-HA. Also, Wilentz, *The Rainy Season*, 272–5.
14 An early 1960s survey found that religious beliefs played a significant role in women's attitudes to family size. (J. Mayone Stycos, "Haitian Attitudes toward Family Size," *Human Organization* 23 (1964): 42–7, cited in Kathleen A. Tobin,

"Population Density and Housing in Port-Au-Prince: Historical Construction of Vulnerability," *Journal of Urban History* 39 (2013): 1051–2.

15 See Schwartz, *Travesty in Haiti:* chapter 6, Kindle.

16 A point made in Asger Leth's film, "Ghosts of Cité Soleil: The Most dangerous Place on Earth" Revolver Entertainment Release, 2008. For a critique of the documentary, see Christopher Garland, "The Rhetoric of Crisis and Foreclosing the Future of Haiti in *Ghosts of Cité Soleil*," in *Haiti and the Americas*, ed. Carla Calarge, et al. (Jackson, MS: University of Mississippi, 2013), 179–98.

17 See Gaston Georges Mérisier, "Etude sur les Enfants Non Scolarisés," Save the Children Report, 2009, cited in Erik Boak, "Study on Governance Challenges for Education in Fragile Situations: Haiti country Report," European Union, 2010.

18 James, *Democratic Insecurities*, 43–5.

19 21% of children aged 5–14 were child labourers during 2002–2011. http://data.un.org/Data.aspx?d=SOWC&f=inID%3a86. Accessed: 15/11/2018. Lindstrom reports around 300,000 involuntary child domestic labourers (*restaveks*) in Haiti. There is also an "industry" in child trafficking (Freedom House, *Countries at the Crossroads 2012-Haiti*, 20 September 2012. Online: www.refworld.org/docid/505c1731c.html. Accessed: 15/11/2018).

20 World Bank and Observatoire National de la Pauvreté et de l'Exclusion Sociale (ONPES), "Investing in People to Fight Poverty in Haiti, Reflections for Evidence-Based Policy Making," Washington, DC, 2014.

21 One aspect of this informal economy is the sex trade, using women's bodies as income for survival. UNHCR *Driven by desperation: Transactional sex as a survival strategy in Port-au-Prince IDP camp*, 2011. Online: www.urd.org/IMG/pdf/SGBV-UNHCR-report2_FINAL.pdf. Accessed 15/11/2018. But violence against women is a concomitant problem: Claire Mcloughlin, "Helpdesk Research Report: Violence against women and girls in Haiti," Governance and Social Development Resource Centre (08/03/2013). Online: http://www.gsdrc.org/docs/open/hdq902.pdf. Accessed: 09/10/2018.

22 Artisans use discarded materials to create "upcycled" pieces of jewellery and home decor. www.apparentproject.org/.

23 *Nàs* is the Kreyol for a basket used for fishing. We saw many instances of *nàs* weaving in the rural communities.

24 Wilentz, *Farewell Fred Voodoo*, 103, 173–5; Casimir, "Two Classes and Two Cultures in Contemporary Haiti." See James, *Democratic Insecurities*, 11. James also indicates developmental policies for decentralisation were being proposed in the 1987 Constitution of Haiti and by the international community prior to the earthquake. (*Democratic Insecurities*, xxi). See also, "Thinking Local in Haiti: Civil Society Perspectives on Decentralisation," in *Report by Progression in Partnership with CAFOD, International HIV-AIDS Alliance, SCIAF and Tearfund* (London: Progressio, 2012), 1–20.

25 S. Blair Hedges, et al., PNAS November 13, 2018 115 (46), 11850–5; published ahead of print October 29, 2018 https://doi.org/10.1073/pnas.1809753115; Wilentz, *The Rainy Season*, 249–54.

26 Jacob Kushner, "US Food Aid Program Is Starving Haiti's Farmers," *The News Herald* Tuesday, January 14, 2014; Katz, *The Big Truck*; Wilentz, *Farewell Fred Voodoo*, 2013; Mark Schuller and Pablo Morales, Tectonic Shifts: Haiti Since the Earthquake, (Sterling, Vir.: Kumarian, 2012); Schwartz, *Travesty in Haiti*.

27 "Haiti-Economy: Everything You Need to Know about the New Minimum Wage," *Haiti Libre*, April 19, 2014; Dan Coughlin and Kim Ives, "WikiLeaks Haiti: Let Them Live on $3 a Day," *The Nation* June 1, 2011; *Haiti Libre*, "Haitian Hemispheric Opportunity through Partnership Encouragement Act of 2006 and 2008," *Federal Register* 73 no. 190 (2008): 56715–29.
28 Wilentz, *Farewell Fred Voodoo*, 255–9; Hallward, *Damming the Flood*, 5–7.
29 Hallward, *Damming the Flood*, 54–65; Wilentz, *The Rainy Season*.
30 James, *Democratic Insecurities*, 68–76.
31 Hallward, *Damming the Flood*, 54–65; Wilentz, *The Rainy Season*, 269–76.
32 Haiti was also subject to the Bumpers Amendment, introduced by U.S. Senator Dale Bumper in 1985. This prevented U.S. Government aid being spent on programmes that could compete with U.S. exports on the global market.
33 Regine Barjon, "Effectiveness of Aid in Haiti and How Private Investment Can Facilitate the Reconstruction," written statement to the U.S. Senate Subcommittees of Foreign Relations on international development and foreign assistance and Western hemisphere in Laurette M. Backer, *Rebuilding Haiti in the Martelly Era*, June, 2011: 7. Online: www.scribd.com/document/67320556/REBUILDING-HAITI-IN-THE-MARTELLY-ERA-June-23-2011. Accessed: 09/11/2015.
34 Farmer, *Haiti after the Earthquake*, 150, and fn.1; Isobel Macdonald, "Clinton Apologizes for U.S. Role in Destroying Haitian Democracy (Happy April Fool's Day!)," *Huffington Post*, January 4, 2011. Online: www.huffingtonpost.com/isabel-macdonald/clinton-apologizes-for-us_b_843541.html. Accessed: 06/06/2014.
35 Jean Casimir, "Two Classes and Two Cultures in Contemporary Haiti," in *Contemporary Caribbean: A Sociological Reader*, vol. 2, ed. Susan Craig (Maracas, Trinidad and Tobago: The College Press, 1982).
36 Anthony Oliver-Smith, "Haiti and the Historical Construction of Disasters," *MACLA Report on the Americas* 43, no. 4 (2010): 32–6. On the impact of neoliberal fiscal policies upon disasters see Naomi Klein, *The Shock Doctrine* (London: Penguin, 2007), and "Disaster capitalism in Action: Haiti," in her blog posts "The Shock Doctrine: The Rise of Disaster Capitalism," January 13, and January 14, 2010 posts. Online: www.naomiklein.org/shock-doctrine/resources/disaster-capitalism-in-action/tags/haiti. Accessed: 05/10/2018; James, *Democratic Insecurities*, 13.
37 Nadeve Menard, "Helping Haiti-Helping Ourselves," in *Haiti Rising: Haitian History, Culture and the Earthquake of 2010*, ed. Martin Munro (Liverpool: Liverpool University Press, 2010), 53.
38 Beverley Bell, "7.0 on the Horror Scale: Notes on the Haitian Earthquake," in Munro, *Haiti Rising*, 163.
39 Marc O. Eberhard, et al. "The m_w 7.0 Haiti Earthquake," USGS / EERI Advance Reconnaissance Team: TEAM REPORT V 1.1, February 23, 2010: 13; Institute Haïtien de Statistique et d'Informatique, Présentation Générale des Resultats: Batiments, *Ministère de L'Economie et des Finances* (2010). Online: www.ihsi.ht/rgph_resultat_ensemble_b.htm. Accessed: 16/11/2018.
40 Eberhard, "The M_w 7.0 Haiti Earthquake," US Geological Survey Open-File Report 2010-10-48. Online: pubs.usgs.gov/of/2010/1048/; Raymond A. Joseph, "Five Years Later: Where Did All the Haiti Aid Go?" *Wall Street Journal*, January 9, 2015. Online: www.wsj.com/articles/raymond-joseph-five-years-later-where-did-all-the-haiti-aid-go-1420847196. Accessed: 09/01/2015.
41 Patrick Paultre, et al., "Damage to Engineered Structures during the 12 January 2010 Haiti (Léogâne) Earthquake," *Canadian Journal of Civil Engineering* 40 (2013): 777–90.

42 See note 41.
43 "Housing Exposition Exposes Waste, Cynicism," *Haiti Grassroots Watch*, September 4, 2012. Online: haitigrassrootswatch.squarespace.com/20eng. Accessed: 28/01/2014. "Haiti: Reconstruction's Housing Projects Still Plagued with Problems Four Years after the Earthquake," *Global Research*, January 13, 2014. Online: www.globalresearch.ca/haiti-reconstructions-housing-projects-still-plagued-with-problems-four-years-after-the-earthquake/5364749. Accessed: 28/01/2014. Christian Werthmann, et al., "Conception du processus: Designing Process." Online: issuu.com/gsdmit/docs/designingprocess.
44 Nicholas Ambraseys and Roger Bilham, "Corruption Kills," *Nature* 469 (2011): 153–5; Roger Bilham, "Societal and Observational Problems in Earthquake Risk Assessments and Their Delivery to Those Most at Risk," *Tectonophysics* 584 (2013): 166–73.
45 Ambraseys and Bilham, "Corruption Kills," 154.
46 Bilham, *Tectonophysics*, 584, 167.
47 Ambraseys and Bilham, "Corruption Kills," 155.
48 On the application of liberation theology in Haiti, see Weiss and Griffin, *In the Company of the Poor*; Sobrino, *Where Is God?* 2–105.

6 What we have to teach the world

Our internet alerts deliver to us a daily diet from faith-based charities, churches and other humanitarian organisations that either want to help Haitians with projects like building orphanages and schools, or to deliver toys or food of some description that people have generously donated. It seems there are many people in High-Income Countries who want to hand on their spare produce, clothing, toys, tools and personal energy, to help the "poor Haitian people" with what they think Haitians need. Many also want to *teach* Haiti, claiming to have the answers to all the ills of "the poorest nation in the Western hemisphere."[1]

The aim of this book has been to permit the Haitian people to speak for themselves. We are persuaded that there are lessons Haitians have to teach those of us who live in countries who do not think "poverty-stricken," "uneducated," "needy" Haitians have anything to teach us. In fact, we are sickened by the levels of hegemonic self-interest, fawning "compassion" and humanitarian patronage to which Haitians are routinely subjected. These are object lessons in how to humiliate a small but noble nation that is, in Ulysse's words, "the new free black Republic that ended slavery and disrupted the order of things in the world [and] became a geopolitical pariah and our humanity was disavowed."[2] Few countries are subjected to so much international bullying as Haiti, while at the same time those same bullies never register how great a debt they owe to Haiti for the political status they can now strut on the world's stage.[3] Even attempts by outsiders acknowledging Haitians' *resilience* – a chic term in disaster studies – can be patronising, as Brunine David explains:

> When they dare to talk about our courage and strength or perseverance, they change the meaning and take all the good from it and leave us with resilience; a kind of people who accept any unacceptable situation, people who can live anywhere in any bad condition that no one else would actually accept.[4]

In this final chapter, we draw together lessons from our survivors' narratives of faith from their experience of the Haiti earthquake disaster for more global contexts of disaster *response* as well as recovery. At the same time, we acknowledge the fact that every disaster has its own unique dynamics. Perhaps one of the clear lessons the Haitians have to teach the world is the folly of applying generic disaster aid and development policies, a point Haitians have learned from being on the receiving end and one to which we shall return more fully below. In particular, we highlight the significant role that religious faith can play in learning, exhibiting and disseminating our lessons through networks of churches and faith-based organisations globally, in a bid to ensure that the human factors of "natural" disasters are addressed with theological and social integrity and ethical equity.

Religion

That religion played significant roles in the responses to and recovery from the Haiti earthquake disaster is beyond dispute. This is because, first, Haiti has a dynamically religious culture, so it is almost impossible to be Haitian without being religious. What is more, being religious, Haitian-style, is to be, usually, vaguely Christian in some form or other, but it is also to be influenced symbiotically and culturally by Haitian Vodou. The Haitian ethnographer Jean Price-Mars captured this relationship when he concluded,

> In any case, not a bud, a breath, a cell can actually escape from the biologic solidarity which binds the living matter to today with the raw energy expended by African negroes with their tears, sweat, and blood in the soil of ancient Quisqueya in order to transform it into our country of Haiti.[5]

The Haiti context teaches the importance of taking into account a clear and close analysis of every cultural context in which a disaster occurs: we conclude that this must take serious account of the theologies and practices of the religious milieu. This lesson was affirmed by Theodore, the Social Services and Mental Health Director, when he said, "If you can talk to any Haitian, you will find in most of them that whatever happened, it was permitted by God . . . this is to me . . . something that people can use to promote resilience. And I think it has helped a lot." This is why Leon could claim,

> It's incredible, but the Haitian nation is a nation full of resistance. And one of the places that did this resistance, this endurance comes from, it is from God because there are a lot of people that believe in God – there are a lot of Haitians that believe in God.

This being said, it does not follow that disaster response agencies, or even a public that is responding to media appeals for aid, will respect such native religious beliefs seriously or un-patronisingly. We conclude that a greater level of attention to, and respect for, religious beliefs in disaster-afflicted contexts is necessary even if responding agents do not hold to, or even have much sympathy with, those beliefs.

Second, it is clear from the selection of narratives we have presented that religious beliefs played a significant role in the way the disaster, in association with participants' pre-existing struggles, was interpreted. These beliefs went a long way to preserving their sanity and their capacity to survive with such measures of hope as they have for the future. To minimise, let alone to ignore or dismiss, the significance of religious beliefs can only serve to undermine a substantial foundation resource for survival and recovery. Religion, carefully nuanced and contextualised, holds a legitimate contribution to disaster response and recovery globally.[6]

Survival strategies

Religious beliefs were by no means the only factors that did play, or that should play, a significant role in catastrophic disaster contexts. Another factor, which played an equally significant role for our Haitian survivors, is learned survival strategies. This warrants highlighting alongside religious beliefs, though, again, religion often plays an attending role. These strategies developed out of, in many cases, routine hardships of Haitian life. The 2010 earthquake came on the heels of an already established discourse of trauma from the previous years of political upheaval and social and infrastructural chaos. Nevertheless, as Erica Caple James confirms, "Supported by a rich cultural heritage, the Haitian people retain a capacity for hope, faith and resilience that remains a tremendous resource for any efforts to rehabilitate the nation and its people."[7] As Delphine explained:

> It has to do with their [Haitian] culture as well. And from the beginning of our history we been through a lot! And they've learned to be strong. They have learned to, sometimes because they have no choice, other than accept what they are living. So, as you said, beside, well, I wouldn't even say Christian, beside being Christian and just being [under] attack. Mentally, it has a lot to do with their culture and then their mental state.

Delphine admitted that some Haitians do too much complaining, then she continued,

> But we do have people with their Christian faith, who have faith in God, and their culture, make them strong from everything they've been

through, they survive them and now they can say, "Well, this can't do much to me, because I've gone through worse," you know? And I think, you know, it's a mental, I think so.

Jean-Francois, a Haitian pastor, told us that Haitians were people who had robust resistance, which came from their strong belief in God. He then spoke about the ironic role the Government of Haiti played in creating survival strategies. He explained:

> We never had a Government that would feed the people that did not work, a Government that would give us good water to drink, or a Government that would give us work for us to do. We never had a Government that gave us clothings, or a Government that give us lively goods so we could live. *And because of those difficulties we have learned how to live by ourselves*. And we can now resist anything that may happen. The country may be sinking, those who have life will rebuild it again. Because we never believe in a Government coming to do something for us. After the earthquake of 2010, we had many difficulties. There were the international community that just plunged upon us, and those that brought food we took because we had problems and would have died had we not found them. Those that came to help us in the medical field

Figure 6.1 Resilience of survivors: wall plaque in Memorial Garden at Hotel Montana.

Source: Photo – Roger P. Abbott

> we embrace them. There are those that said they'd bring houses, but though that was not true, those that they did bring we took. And so with all these things we have taken; now the international NGOs institutions, they have left and now we are left on our own *and never once did we consider that the international aid that came that they would help us nor the Government that came into power, though they had good visions for the country. But we never once believed that they would be able to provide . . . they would be able to do something for us. After the earthquake we have a cyclone, we have cholera, and political crisis, but a month even after all this we have resisted, we have resisted because we are a nation that does not believe in the Government doing something for us nor an international or a foreign aid doing it, OK, because we believe in our own hope, that we can do something for ourselves.* And if the earthquake had happened in a foreign country the nations that were victims, they would just stand there with their arms crossed. Why? Because their country has disaster relief plans, they have social welfares, and the country would take care of them, but us, if we have only one sweet potato, we will make it last for the amount of days it should last. *Because—and that* [is] *one of the things that help us a lot, that resistance—because as a nation we do not have any place where we could deposit our problems and expect any solutions.*
>
> (Emphasis ours)

Most Haitian hardships have their roots in the histories of struggle in Haiti, and not least in the tap-roots of slavery and the battles against foreign colonisation and for independence. These well-documented battles and the bitter, unforgiving and punitive tastes they have left in former foreign colonisers' cultural and political memories are crucial to understanding the Haitian zeitgeist of today. Failure, wilful or otherwise, to take these historical realities seriously, in particular any sense of "other nations" complicity in contributing to the vulnerability of Haiti when natural hazards strike, has played a large role in the failure of INGOs and international communities to bring adequate recovery and development to this catastrophised country even ten years after the earthquake. In particular, we believe the international Christian community needs to play a much more proactive part in acknowledging the horrible role she has played historically, in sanctioning and "blessing" the "Middle Passage" of the slave trade. Those who survived that passage were then baptised in the name of the Holy Trinity within eight days of their arrival before being de-humanised and commoditised into a life of slavery in order to help the colonisers exploit and strip the resource, and strategically rich "Pearl of the Antilles." This deep scar in the Haitian psyche means that Christianity's attempts at bringing "solutions" for the country's ills and her future development, in our view, must be steeped in humility and in a rejection of the historic racism and colonial hegemony that inflicted the original wound and has kept it open. The contribution of Vodou as a cultural dynamic element to generational survival strategies cannot be ignored, minimised or demonised, if an honest account of those strategies is to be recognised.

Education

Those populations that are most vulnerable to seismic hazards are the poor, who often lack access to good education. They are the most likely to die, they are often left with life-changing injuries, and their livelihoods are the most likely to be destroyed. Often they also happen to be deeply religious populations. Our research shows that there is a real hunger for education in disaster mitigation among vulnerable communities hampered by poverty. We found that poverty-restricted access to education does not equate to poverty of interest in seismic risk or to any lack of desire to be proactive in taking mitigation measures: indeed, quite the reverse.

Our Haitian survivors, in addition to demonstrating the value of religious belief, also tell us how essential the scientific research needs to be. For example, at The Faraday Institute for Science and Religion, Cambridge,

Figure 6.2 Education: most schools in Haiti, including this one, are faith-based.

Source: Photo – Roger P. Abbott

U.K., we explore ways for theologians, faith communities and geoscientists to collaborate in the interests of risk reduction and mitigation in areas of natural hazard vulnerability. Often geoscientific and religious communities have tended to mutually ignore the vital roles each can play in collaboration in the best interests for saving lives as well as for saving economic livelihoods that can lead people out of poverty and provide better access to education. The geoscientist can provide clear scientific, evidence-based contributions for educating vulnerable communities about their particular seismic hazards, their degrees of risk and possible actions for disaster mitigation. On the other hand, the practical theologian can contribute a pastorally and culturally sensitised case for encouraging and assuring those communities that they can take the science seriously, and for assuaging any doubts or conflicts about geoscience and religion being incompatible disciplines. Any theology/religion around natural hazards and disasters that ignores or denigrates the contribution of hard science is a seriously dysfunctional, defective and dangerous religion!

Given the rapid expansion of scientific knowledge of natural hazards, including earthquake hazards, we commend the moves being made recently by the Government of Haiti to access and develop relevant scientific research in natural hazard risk mitigation. However, we recommend that priority be given to ensuring that populations in natural hazard regions urgently receive the education necessary to be aware of the nature of the hazards and of how to mitigate risks to life and living; in particular we recommend the developments being made in smart technology.[8] We recognise, however, that for this to happen soon then the international community will have to provide financial and resource assistance, targeted to these specific aims.

The scientific education should also be contextualised and expanded to provide guidance that includes identifying safe zoning and methods for hazard-resistant building, with the sensitive enforcement of regulations. The proverbial, "Earthquakes don't kill, buildings do," should be emphasised among civic communities, as should access to safe, affordable housing and, if necessary, subsidised safe housing for the low-waged and unemployed.

We recommend that there should be regular disaster exercises and rehearsals for communities, as are common in other earthquake-prone countries. Our research shows the importance of the role faith communities can play, since they often contain strategic, influential community leaders and disseminators, as well as educators. When armed with an indigenised civic theology and compassion, faith communities could take a significant role in proactive disaster awareness and rehearsals. What can often prevent this kind of thing happening, as we have seen in Haiti and in other disaster-prone regions, is either strong resistance by central and/or local governments to the role of faith communities as such, or strong resistance by the faith communities to becoming "chartered by the Emperor," as the theological ethicist Stanley Hauerwas describes.[9] We suggest that Haitian faith communities, particularly some of the Evangelical Protestant communities, should be challenged over their reluctance to participate and respond ecumenically in the interests of social cohesion in disaster awareness, planning and response. This reluctance becomes acute when the influences of Vodou are taken into account. We believe faith communities could serve the interests of social cohesion in times of national crisis, which do not need to be interpreted as compromising their faith or require practices that are unfaithful to theology or liturgy.[10]

Governance

Another significant factor is the issue of political governance (State) for the people (Nation).[11] Governments need to take the science seriously and to take appropriate actions. Kathleen Tierney, Director of the Natural Hazards Center at the Institute of Behavioral Science in Colorado, made the point when she said,

> We already know a great deal of what we need to know in order to reduce the pain, suffering and other losses associated with disasters, but . . . applying that knowledge is difficult because of institutional inertia and especially because of the benefits those in power obtain through activities that increase that risk.[12]

Paul Farmer highlights another important dimension to the problem: "The history that turned the world's wealthiest slave colony into the hemisphere's poorest country has been tough, in part because of a lack of respect for democracy both among Haiti's small elite and in successive French and U.S. governments."[13] The culture of corruption endemic among the elite and political classes in Haiti, and the interference from foreign governments, has played an unconscionable part in depriving Haitian citizens of life-saving education about hazard awareness and disaster risk reduction measures. It is clear from rural narratives that citizens in Haiti feel politically isolated from the State executive, the Government of Haiti, or, as some Haitians refer to it, the "Government of Port-au-Prince." Therefore, the provincial and rural populations suffer even more from lack of education and of resources.

It is not, however, only the Haitian elite and political class that need to adjust their view of the well-being of Haitian citizens. So does the international community. As Haitian Westenley Alcenat has commented,

> And yet, the global response has been the same as usual: rather than examine how the complex intersections of history, politics, economics

> and ecology conspire to make Haiti susceptible to natural disasters and epidemics, journalists, pundits, and NGO operatives instead shift blame onto Haitians themselves. They present Haitians as a people incapable of managing their nation. This view has guided the international response to Haiti since its independence two hundred years ago.[14]

As we have indicated earlier, there are those who consider that what Haitians need do is to repent of pacts with the devil, that being the nation's "original sin," because of which it has never been allowed by divine providence to prosper and under which the country has suffered a perpetual curse. We do not share that view. We concur with the former Organisation of American States' Special Representative in Haiti, Ricardo Seitenfus' view that, "Haiti's original sin, on the world stage, was its liberation . . ." because, "the revolutionary model scared the colonialist and racist Great Powers." Seitenfus' premature dismissal from office in early 2011 soon after making his views known in public reveals how much his rejection of UN, OAS and NGO policy for helping earthquake- and cholera-decimated Haiti had registered on the international Richter Scale for annoyance. We believe he was right and honourable in adopting the whistle-blower stance he did, even at such personal cost. His point, when he stated, "And above all, we have to think that the development of Haiti has to be done by Haitians. If you imagine you can do this through MINUSTAH and through NGOs, we will be deceiving Haitians and misleading world public opinion" is one that holds key importance for all NGOs involved in disaster response and relief. Indeed, it has echoes of an earlier comment made by Price-Mars, that "NGOism amounts to 'imperialism of every order [that] disguises its lusts under the appearance of philanthropy.'"[15]

We remind the INGO community that during a time of emergency they are guests in the country they enter to give disaster assistance. They should avoid giving any impression that they are occupiers or invaders. Respect for national sovereignty in times of crisis is crucial to enable a disaster-afflicted country to recover and develop. The necessity for strong and fair State-governance, and the end of "the real or imagined transactions of magic or sorcery for material or political gain," is crucial in enabling this disaster-afflicted country to become a safer place to live in light of the natural hazards to which she is exposed.[16]

Poverty is one of the main drivers in causing natural hazards to develop into disasters. In 2015, the leaders of nineteen governments including Haiti promised to reduce inequality in their nations as Goal 10 of the Sustainable Development Goals (SDGs), in a move toward eradicating global poverty. In 2017, Development Finance International (DFI) and Oxfam produced an index for measuring these promises to reduce the gap between the rich and the poor. The 2018 edition of the DFI and Oxfam Report placed Haiti at 155th out of 157 countries measured. The criterion for measuring were: social care in the form of universal, public and free education and universal social protection floors; that commitments should be funded by increasing taxation and clamping down on tax dodging; the country must commit to respecting workers' rights and there should be a raised level of minimum working wage. The best Haiti could achieve was in spending on education, health and social protection (133rd), the worst measure was in labour and minimum wages (156th).[17] These scores make for stark reading and throw a light on the way Haiti is influenced so much by a very small minority of elite families, who dominate commerce and governance and who exert a stranglehold on workers' rights and wages. These people flourish from maintaining the vast majority of the population of Haiti in the structural sin of poverty.[18] Yet even at the back of these problems lies the even stronger historic influence of debt-enforcement that foreign governments have maintained on the country's Treasury in the past, thus hampering any realistic fiscal capacity to uplift itself towards attaining Goal 10.[19]

Theology and structural transformation

History, not geography or geology, accounts for Haiti's vulnerability to natural hazards;[20] the only aspect of nature that is actually problematic is *human* nature, all other aspects are innocent but are suffering (Romans 8:20–21). The seismic and meteorological hazards Haiti experiences are natural phenomena of creation, and they should be accounted for as such, not just geophysically but also theologically. To insist that disasters, such as happen in Haiti, are natural, as the common evangelical hermeneutic too often does, in theological terms must mean that they are the products of divinely created natural evil, and, therefore, on this basis we must insist that God is responsible for the disasters that ensue.[21] Continuing this evangelical hermeneutic for explaining hazards, such as earthquakes and storms, as natural evils is incorrect and misleading scientifically and theologically. What is more, the view feeds the conviction that such events cannot be prevented; they are inevitable since even the natural world is now under the divine curse (Genesis 3:17), and so we just have to live with the fact that disasters happen, and they happen because of the natural evil God creates. We reject that view, since the Bible teaches that it is not the works of creation that represent the heart of the problem for humans; it is we humans ourselves. The "ground is cursed" *because of us*! (Genesis 3:17). In other words, it is human sin and evil that accounts for the way the necessary wonders of creation, such as earthquakes and storms, become disastrous to humans. God has exposed his good creation to the abuses of human sin; hence, the creation "groans" (Romans 8:22).

It is clear from historic and contemporary life that earthquakes do not have to injure or kill. Even within the severe limitations of their "fallen" nature, humans have been able to find ways to live in relative peace and safety with natural hazards. Those ways have failed when some human factors have been at play in the run-up to and during the hazard event.

For this reason, we aver that keeping in vogue the common evangelical view plays into endorsing a passivity and inertia regarding the very obvious and appalling human errors and evils that turn natural hazards into disasters. We recommend that the term "*natural* disaster" be dropped from the researcher's lexicon, so that it may also die out as the dishonest term it is in religious, public and media discourse on disasters. Secular geographers and social scientists are ahead of the game compared with Christian theologians, religious leaders and the media. The following comments, with which in considerable measure we agree, give some indication of secular perspectives:

> In short, disasters are not accidents or acts of God. They are deeply rooted in the social, economic, and environmental history of the societies where they occur. Moreover, disasters are far more than catastrophic events; they are processes that unfold through time, and their causes are deeply embedded in societal history. As such, disasters have historical roots, unfolding presents, and potential futures according to the forms of reconstruction. In effect, a disaster is made inevitable by the historically produced pattern of vulnerability, evidenced in the location, infrastructure, sociopolitical structure, production patterns, and ideology that characterizes a society.[22]

and,

> It is generally accepted among environmental geographers that there is no such thing as a natural disaster. In every phase and aspect of a disaster – causes, vulnerability, preparedness, results and response, and reconstruction – the contours of disaster and the difference between who lives and who dies is to a greater or lesser extent a social calculus. Hurricane Katrina provides the most startling confirmation of that axiom. This is not simply an academic point but a practical one, and it has everything to do with how societies prepare for and absorb natural events and how they can or should reconstruct afterward.[23]

A group of scholars who are engaged in disaster risk reduction produced a guidance note for the media in 2017, encouraging journalists and reporters to ditch using the label, "natural disaster." In their view, the label is "incorrect and misleading," and "the attribution of blame for disaster losses to nature, or to an 'act of God,' absolves powerful decision-makers of responsibility for allowing or forcing people to live in vulnerable conditions," and "This use of language strips disasters of their social, political, environmental and economic context – one where injustice is pervasive."[24] As a theologian and a seismologist, we endorse these opinions and the campaign for the term to be dropped from both academic research and public vocabularies for describing natural hazard events that become disasters. Doing so can only help allocate the blame where it ought to lie, on human causal factors, enhancing the message, both in the academic and public spheres, that disasters can be mitigated if people are willing to address these factors. This is also the reason one of us entitled their book on the science and theology of disasters as *Who Is to Blame: Nature, Disasters and Acts of God* rather than *Natural Disasters*.[25] The Haiti earthquake is the *cause celebre* for demonstrating the validity of this point.

As evangelical Christian authors we are clear that human *evil* is what turns natural hazards into disasters. Additionally, we align with Donald Gelpi's call for a socio-political conversion. This expands a typical evangelical focus upon personal sin and individual atonement toward the insistence that "the Christian faithful acknowledge there are terribly sinful consequences deeply embedded in the way our prevailing social systems operate, and that the Christian life mandates solidarity with the poor through social and political action."[26]

However, an important caveat is required at this point. When we say that evil is what turns natural hazards into disasters we are not implying that disasters exclusively target the guilty. If this were the case, then why would it be that it is the poorest and those who lack the life-saving education and material resources, who suffer and die the most, including multitudes of young children, mothers and the elderly – "innocents" – while the financially and resource-rich, who usually live in hazard-safe regions, do not?

The kind of evils that turn natural hazards into disasters, such as happened so obviously in Haiti, are what we call structural evils. David Chester has drawn attention to this, concurring with the work by those who lean toward a (non-Marxian) version of liberation theology.[27] The version of liberation theology these scholars favour is that developed by Fr. Gustavo Gutiérrez, and is also the theological approach to Haiti's problems now espoused and championed by Paul Farmer.[28] Farmer, having lived and worked in Haiti for many years prior to as well as since the earthquake, has an astute and indigenised awareness of the root problems Haiti has, not least in the field of healthcare and the dark intercourse between U.S. and Haitian politics.[29]

What is particularly attractive, theologically and ethically, in the context of Haiti's future, is the principle Gutiérrez inspired in Farmer's commitment to healthcare in Haiti, under the *Zanme Lasante* (Partners in Health) organisation. That principle was a preferential option for the poor.[30] Application of this principle requires four commitments: to the poor for life; to raising prophetic voices in the public square; to praxis and theory; and to extending the Kingdom of God in the here and now.[31] Where Farmer observed Haitians becoming "socialised for scarcity and failure,"[32] we observe they are also socialised for *disaster* and failure, in terms of levels of expectation from the structural systems maintained by the elites and politicians.

In view of the common controversial tendency to link liberation theology with Marxism, we would suggest that it is possible to place the kind of theological *praxis* espoused by Guttierez and Farmer into the category of practical theology.[33] It is clear that little will change in Haiti regarding the social, political and commercial institutions and structures that currently conspire, and collaborate, in order to deprive the population of education and resources for disaster risk reduction, unless there is real change within these structures.[34]

There are precedents for working for structural change in Haiti through religious communities. In the late 1970s, Fr. Jean-Paul Aristide, motivated by his liberation theology, saw the *ti kominote legliz* (the church community) and *ti legliz* (small church) movements emerge.[35] The local ecclesial communities became instrumental in organising protests seeking justice over the offences of the Duvalierist *Ton Ton Macout*.[36] Unfortunately, such grassroots movements met with violent anti-Aristide repression by the military and elite classes in Haiti at that time, and there is no guarantee that a *ti legliz* today would not suffer similar reactions against them from the elites. However, exploring a model for such grassroots faith-based actors who have a selfless passion for civic safety and who are sick of violence, conflict and national humiliation in their natural hazard vulnerable nation, could become a model worth exploring and testing. We conclude that what Haiti needs most is not the charity/aid-based solution it has laboured under for too long, since this only creates need to perpetuate the industry.[37] According to our participants, Haiti requires profound change to eradicate the structural evil of poverty. Our participants did not wish us to just tell the story of their sufferings, but also to tell the need for structural change in Haiti. Christina Zarowsky, as she reported Somalian refugees' experiences of trauma, made a similar request:

> They did not wish me to stop at conveying their individual misery, for they knew it well enough and did not consider that emotional empathy was sufficient to resolving their difficulties. . . . If this insistence on building a politicized collective memory and master narrative challenging power and injustice from the local to the global represents "trauma,"

> it is of a different scope and implies different therapeutic interventions than those suggested by conventional models of PTSD.[38]

For our own participants, Haiti requires "politicians with national agendas, not self-interest, one that recognizes its duty to its citizens."[39] Charity and especially solidarity have their place, but above all, "respecting the status of the poor as those who control their own destiny is an indispensable condition for genuine solidarity."[40] We agree with Gutiérrez, therefore, that the solution for Haiti needs to be theological as well as political.[41] As if to reinforce this point, on the evening of October 6th, 2018, a magnitude 5.9 earthquake struck the north-east and Artibonite regions of Haiti, resulting in eighteen deaths and 333 persons injured. Moreover, 7,430 houses were destroyed. The scale of deaths, injuries and damage to houses was out of proportion for an earthquake of this mid-range magnitude. The Port-au-Prince daily newspaper reported,

> The feeling of panic that seized every Haitian who felt the tremors, all over the country, and the deprivation of the institutions of Port-au-Prince as province showed that there is still work to reach the excellence in disaster preparedness like an earthquake.[42]

Claude Prepetit lamented, "If a magnitude 5.9 earthquake can do so much damage, imagine for a moment that the magnitude was the one we knew on January 12, 2010." We rest our case!

As this book goes to the publisher, Haiti is experiencing a week of peace, relative to the "Operation Lockdown" over the previous three weeks in February 2019, which had seen the closure of banks, commercial businesses, schools and churches, and which placed enormous strains on healthcare. "Operation Lockdown" came about from grassroots mass protests by the Haitian populace over issues going back to previous riots that took place in October of 2018 against the State and the elite's collusion in corruption over the Petro Caribe debacle involving a Venezuela-sponsored oil assistance project, and the "mysterious" disappearance of funds earmarked under the project for use for improving social, educational and public services in Haiti. Such protests then mushroomed into violence, under the addition of various politically motivated parties, into a demand for the resignations of the current President, Jovenel Moise, and his Prime Minister, Jean Henry Ceant, on account of their alleged involvement in the Petro Caribe corruption affair and in the general financial crisis the country currently faces.

These recent events are the outcome of a number of the social justice factors mentioned in this book and indicate the seriousness of the precarious State-Nation relationship in Haiti, and the population's extreme vulnerability

The damaged Presidential Palace, Port-au-Prince.

Source: robertharding/Christian Kober

to a catastrophic outcome in the event of another earthquake, or indeed of any extreme major natural hazard in the immediate future. God forbid any such occurrence will happen, but should such a hazard occur during these times, we trust the real question that will be raised will not be, "Why did God allow such suffering to occur?" but rather, "Why did humans allow it?"

Notes

1 Wilentz, *Farewell Fred Voodoo*, 274–6, 301–8.
2 Ulysses, *Haiti Needs New Narratives*, 60.
3 Calargé, et al., *Haiti and the Americas*; Naomi Klein, "Haiti: A Creditor, Not a Debtor," 2010. www.naomiklein .org/articles/2010/02/haiti-creditor-not-debtor.
4 Quoted in Ulysse, *Haiti Needs New Narratives*, 61.
5 Price-Mars, *So Spoke the Uncle*, 104.
6 Abbott, *Sit on Our Hands*, 129–31, 363–71.
7 James, *Democratic Insecurities*, xxiv; also Alex Dupey, "Commentary Beyond the Earthquake: A Wake-Up Call for Haiti," *Latin American Perspectives* 37 (May 2010): 195–204. doi: 10.1177/0094582X10366539.
8 A visit by a group from the United States Geological Survey to Haiti in 2018 ended with announcing development of a national awareness programme to reduce loss of life and property during an earthquake, and the addition of two seismological stations, making twelve in all. ("Haiti-Security: Seismic Risks,

Mission of American Experts in Haiti," *Haiti Libre*, 26/20/2018). On the contribution of technology, see Trevor Nace, "How Technology Is Advancing Emergency Response and Survival during Natural Disasters," *Forbes*, December 15, 2017. Online: www.forbes.com/sites/trevornace/2017/12/15/how-technology-is-advancing-emergency-response-and-survival-during-natural-disasters/. Accessed: 20/12/2018; C. White, *Social Media, Crisis Communication, and Emergency Management* (Boca Raton: CRC, 2012); Tim Large, "Cell Phones and Radios Help Save Lives after Haiti Earthquake," *Reuters*, January 25, 2010. Online: www.reuters.com/article/us-haiti-telecoms/cell-phones-and-radios-help-save-lives-after-haiti-earthquake-idUSTRE60O07M20100125?sp=true. Accessed: 21/12/2018.

9 Stanley Hauerwas and William H. Willimon, *Resident Aliens: Life in the Christian Colony* (Nashville: Abingdon, 1996), 39.

10 Abbott, *Sit on Our Hands*, 309–60.

11 Trouillot, *Haiti*, 19–24.

12 Kathleen Tierney, *The Social Roots of Risk: Producing Disasters, Promoting Resilience* (Stanford, CA: Stanford University Press, 2013): 215–42, in John McClure, "Fatalism, Causal Reasoning, and Natural Hazards," *Natural Hazard Science, Oxford Research Encyclopedias* (2017). Online: doi: 10.1093/acrefore/9780199389407.013.39.

13 Paul Farmer, "Haiti's Unnatural Disaster." *The Nation*, September 17, 2008. Online: www.thenation.com/article/haitis-unnatural-disaster. Accessed: 07/06/2014.

14 Alcenat, "The Case for Haitian Reparations."

15 Price-Mars, *So Says the Uncle*, 10.

16 James, *Democratic Insecurities*, 30.

17 "The Commitment to Recuing Inequality Index 2018: A Global Ranking of Governments Based on What They Are Doing to Tackle the Gap between Rich and Poor," A Report by the Development Finance International and Oxfam (2018): 53.

18 Griffin and Weiss, *In the Company of the Poor*, 16, 131–5; Justin Thacker, *Global Poverty: A Theological Guide* (London: SCM, 2017), 79–91.

19 Jubilee Debt Campaign (2008). Online: web.archive.org/web/20100120065057/www.jubileedebtcampaign.org.uk/Haiti+3113.twl In 2015 France waved c. $77 million in modern debt but also refused to repay the historic compensation she had demanded from Haiti following the revolution and independence of 1804, a sum of around $21 billion in today's money. President Aristide had demanded repayment to Haiti back in 2004.

20 Farmer, *Haiti after the Earthquake*, 2–5.

21 The evangelical view has been described, in a rather enigmatic attempt, by the evangelist/apologist John Blanchard, in his *Does God Believe in Atheists?* (Darlington: Evangelical Press, 2000), 537. It is enigmatic because he had previously acknowledged human causation of many disasters (533–4).

22 Anthony Oliver-Smith, *"Haiti and the Historical Construction of Disasters," NACLA*, July 2010: n.p. Online: nacla.org/article/haiti-and-historical-construction-disasters.

23 Neil Smith, "There's No Such Thing as a Natural Disaster," June 11, 2006. Online: understandingkatrina.ssrc org/Smith/. Accessed: 08/02/2018.

24 Kevin Blanchard, "#NoNaturalDisasters: Changing the Discourse of Disaster Reporting," DRR Dynamics: Supporting Inclusive and Responsible Disaster Risk Reduction (2017). Online: https://docs.google.com/document/d/1wauOb1qVH3ID_YZOzkawhCYOROcybxaD3vIMg3DFNUs/edit. Accessed: 22/10/2018.

25 Robert S. White, *Who Is to Blame? Nature, Disasters and Acts of God* (Oxford: Lion Hudson, 2014), 207. ISBN 978-0-85721-4737.

26 Griffin and Weiss, *In the Company of the Poor*, 2; see also the *observe, judge, act* methodology Farmer explains in *In the Company of the Poor*, 37–44.

27 Sobrino, *Where Is God*, xxxi, 82; David Chester, "Natural Disasters and Christian Theology," The Faraday Institute for Science and Religion. Online: faraday-institute.org/resources/FAR268%20Chester.pdf

28 David Chester, "The Theodicy of Natural Disasters," *Scottish Journal of Theology* 51, no. 4 (1998): 485–505; G. Gutierrez, *A Theology of Liberation* (Maryknoll, NY: Orbis Books, 1988).

29 See Griffin and Weiss, *In the Company of the Poor*; Farmer, *Haiti after the Earthquake*.

30 Gustavo Gutiérrez, *We Drank from Our Own Wells: The Spiritual Journey of a People*, (London: SCM, 1984).

31 Griffin and Weiss, *In the Company of the Poor*, 147–59.

32 Griffin and Weiss, *In the Company of the Poor*, 16.

33 For example, see how Ray Anderson conjoins liberation theology into his practical theology, in Ray S. Anderson, *The Shape of Practical Theology: Empowering Ministry with Theological Praxis* (Downers Grove, IL: Inter Varsity Press, 2001), 102–12; also Griffin and Weiss, *In the Company of the Poor*, 165–6.

34 Farmer, *Haiti after the Earthquake*, 121–39.

35 Hallward, *Damming the Flood*, 155–6; Wilentz, *The Rainy Season*, 105–6.

36 J. B. Aristide and A. Wilentz, *In the Parish of the Poor: Writings from Haiti* (Maryknoll, NY: Orbis Books, 1990).

37 Thacker, *Global Poverty*, 205.

38 Christina Zarowsky, "Writing Trauma: Emotion, Ethnography, and the Politics of Suffering among Somali Returnees in Ethiopia," *Culture, Medicine and Psychiatry* 28 (2004): 189.

39 Ulysse, *Haiti Needs New Narratives*, 8. We commend Thacker's, *Global Poverty*.

40 Griffin and Weiss, *In the Company of the Poor*, 156.

41 Gutiérrez, *We Drank from Our Own Wells*, 50–1.

42 "Haiti, Weak against the Smallest Catastrophe . . . " and "An Earthquake of Magnitude 5.9 Should Not Cause as Much Damage, According to Claude Prépetit," *Le Neuvelliste*, October 8, 2018.

Select bibliography

Abbott, Roger Philip. *Sit on Our Hands or Stand on Our Feet? Exploring a Practical Theology of Major Incident Response for the Evangelical Catholic Christian Church in the UK*. Eugene: OR: Wipf & Stock, 2013.

Calargé, Carla, Raphael Dalleo, Luis Duno-Gottberg, and Clevis Headlet, eds. *Haiti and the Americas*. Jackson, MS: University of Mississippi, 2013.

Farmer, Paul. *After the Earthquake*. New York: Public Affairs, 2011 edition.

Farmer, Paul. *AIDS and Accusation: Haiti and the Geography of Blame*. London: University of California, 2006.

Griffin, Michael, and Jennie Weiss Block. *In the Company of the Poor: Conversations with Dr. Paul Farmer and Fr. Gustavo Guttiérrez*. Maryknoll, NY: Orbis, 2013.

Hallward, Peter. *Damming the Flood: Haiti and the Politics of Containment*. London: Verso, 2010 edition.

James, Erica Caple. *Domestic Insecurities: Violence, Trauma, and Intervention in Haiti*. London: University of California Press, 2010.

Kathleen Tierney, *The social roots of risk: Producing disasters, promoting resilience*, (Stanford, CA: Stanford University Press, 2013): 215–42, cited in John McClure, "Mitigation, Resilience, Risk Communication and Warnings, Climate Change," Preparedness Online Publication Date: April 2017 DOI: 10.1093/acrefore/9780199389407.013.3.

Katz, Jonathan. *The Big Truck That Went By: How the World Came to Save Haiti and Left Behind a Disaster*. London: Palgrave Macmillan, 2013.

Kidder, Tracey. *Mountains Beyond Mountains: One Doctor's Quest to Heal the World*. London: Profile Books, 2009.

Klein, Naomi. *The Shock Doctrine*. London: Penguin, 2007.

Munroe Martin, ed. *Haiti Rising: Haitian History, Culture, and the Earthquake of 2010*. Liverpool: Liverpool University Press, 2010.

Patrick Paultre, Éric Calais, Jean Proulx, Claude Prépetit, Steeve Ambroise, "Damage to engineered structures during the 12 January 2010 Haiti (Léogâne) earthquake, *Canadian Journal of Civil Engineering*, 40 (2013): 777–90. https://doi.org/10.1139/cjce-2012-0247.

Price-Mars, Jean. *So Spoke the Uncle*. Washington, DC: Three Continents, 1983 edition.

Sobrino, Jon. *Where Is God? Earthquake, Terrorism, Barbarity*. Translated by Margaret Wilde. New York: Orbis, 2006.

Thacker, Justin. *Global Poverty: A Theological Guide*. London: SCM, 2017.

White, Robert S. *Who Is to Blame? Disasters, Nature and Acts of God*. Oxford: Lion Hudson, 2014.

Wilentz, Amy. *Farewell Fred Voodoo: A Letter from Haiti*. New York: Simon & Schuster, 2013.

Wilentz, Amy. *The Rainy Season: Haiti Since Duvalier*. New York: Simon & Schuster, 1989.

William Jr. Pierre-Louis, "Haiti Carnival: Misplaced Priorities?" *The Haitian Times*, Feb. 23, 2019. Online: https://haitiantimes.com/2019/02/25/haiti-carnival-misplaced-priorities/. Accessed: 23/02/2019.

Author/Name Index

Subject Index